Integrated Pest Management: Principles and Practice

Integrated Pest Management: Principles and Practice

Walter Ramon

SYRAWOOD
PUBLISHING HOUSE
New York

Published by Syrawood Publishing House,
750 Third Avenue, 9th Floor,
New York, NY 10017, USA
www.syrawoodpublishinghouse.com

Integrated Pest Management: Principles and Practice
Walter Ramon

International Standard Book Number: 978-1-68286-830-0 (Hardback)

Cataloging-in-Publication Data

Integrated pest management : principles and practice / Walter Ramon.
p. cm.
Includes bibliographical references and index.
ISBN 978-1-68286-830-0
1. Pests--Integrated control. 2. Agricultural pests--Integrated control.
I. Ramon, Walter.
SB950 .I58 2019
632.9--dc23

Printed in China.

TABLE OF CONTENTS

PREFACE

Integrated pest management (IPM) is an innovative approach to the economic control of pests. Reducing pest populations below the economic injury level is the chief objective of IPM. It involves various pest control techniques that besides keeping pest populations in check also keep pesticides and other pest management interventions to economically justifiable levels and reduce the risk to human health and the environment. IPM lays emphasis on natural pest control mechanisms with minimum disruption to agro-ecosystems. It has applications in agriculture, horticulture, preventive conservation, turf pest management, ornamental pest management and preventive conservation, besides others. This book elucidates the concepts and innovative models around prospective developments with respect to integrated pest management. It aims to shed light on some of the unexplored aspects of this field. Coherent flow of topics, student-friendly language and extensive use of examples make this book an invaluable source of knowledge.

To facilitate a deeper understanding of the contents of this book a short introduction of every chapter is written below:

Chapter 1- Integrated pest management is an approach to the economic control of pests which integrates a number of pest control techniques. This is an introductory chapter, which will introduce briefly all the significant aspects of pest control such as integrated pest control management and push-pull agricultural pest management.

Chapter 2- A pest is an organism that is detrimental to the well being of humans, agriculture, livestock and forestry. Pests can be of two kinds, vertebrate and invertebrate. This chapter discusses in detail the different kinds of invertebrate pests, such as cereal leaf beetle, ground beetle, boll weevil, Colorado potato beetle, Japanese beetle, etc. as well as vertebrate pests like common myna, pigeon pests, rodent, mole and rabbits. It also elucidates the damage caused by such pests and the varied mechanisms to control their impact.

Chapter 3- The methods that are used to deter or eliminate pest birds from nesting, roosting and landing fall under bird control. This chapter closely examines the varied ways to ensure bird control, such as by using bird control spike, bird netting, avicide, bird scarer and scarecrow.

Chapter 4- Pesticides are chemical substances or biological agents used for the control of pests and weeds. Depending on the pests targeted, they can be classified as herbicides, molluscicides, acaricides, piscicides and dicofol. This chapter includes detailed information about such common pesticides and their role in pest control and management.

Chapter 5- Insecticides are chemicals that are used to kill insects in agriculture and in industry. They are generally distinguished into three types, such as natural, organic and inorganic insecticides. This chapter has been carefully written to provide an easy understanding of the varied types of insecticides and their role in pest control and management.

I would like to share the credit of this book with my editorial team who worked tirelessly on this book. I owe the completion of this book to the never-ending support of my family, who supported me throughout the project.

Walter Ramon

Chapter 1

An Introduction to Pest Control and Integrated Pest Management

Integrated pest management is an approach to the economic control of pests, which integrate a number of pest control techniques. This is an introductory chapter, which will introduce briefly all the significant aspects of pest control such as integrated pest control management and push-pull agricultural pest management.

Pest Control

Pest control is a huge category in the horticulture industry. Essentially, pest control encompasses everything a gardener does both to prevent pests from attacking their plants, and eradicate pest infestations once they occur.

In horticulture, especially in greenhouses and grow rooms, pest control options include chemical and organic pesticides, as well as a host of other options such as integrated pest management (IPM), sticky traps, screens, good personal hygiene, environmental control, and more.

There are several branches of pest control, including biological pest control, which is pitting nature against nature. In these cases, beneficial insects are unleashed in a confined space to attack (eat) the pest insects.

Pest control is an essential part of agriculture. Indeed, pests can reduce food security, having a severe impact on yields. While pesticides are commonly used to treat various types of pests, these substances have also been shown to be quite harmful to plants, in some cases even killing them. An abuse (overuse) of pesticides can even pollute water supplies while having a negative impact on the overall health of the farmers.

For the above reasons, some farmers opt for alternative pest control methods such as crop rotation. This involves occasionally alternating between different crop species in order to stop the pests from getting used to the same types of plants. This method also has the advantage of encouraging soil fertility through nitrogen diffused from the roots.

Alternatively, some planters also choose to go for polyculture instead of pesticides. This entails growing different types of crops in the same space, which has been known to increase local biodiversity while reducing the crop's susceptibility to different types of pests and diseases. On the flip side, polyculture does require more manpower and maintenance.

Other viable pest control options include setting sticky traps up, or planting banker plants, which, in addition to polyculture, is a form of companion planting.

Control Methods

Biological Pest Control

Biological pest control: parasitoid wasp (*Cotesia congregata)* adult with pupal cocoons on its host, a tobacco hornworm *Manduca sexta* (green background)

Biological pest control is a method of controlling pests such as insects and mites by using other organisms. It relies on predation, parasitism, herbivory or other natural mechanisms, but typically also involves an active human management role. Classical biological control involves the introduction of natural enemies of the pest that are bred in the laboratory and released into the environment. An alternative approach is to augment the natural enemies that occur in a particular area by releasing more, either in small, repeated batches, or in a single large-scale release. Ideally, the released organism will breed and survive, and provide long-term control. Biological control can be an important component of an integrated pest management programme.

For example: mosquitoes are often controlled by putting Bt Bacillus thuringiensis ssp. israelensis, a bacterium that infects and kills mosquito larvae, in local water sources.

Cultural Control

Mechanical pest control is the use of hands-on techniques as well as simple equipment and devices, that provides a protective barrier between plants and insects. This is referred to as tillage and is one of the oldest methods of weed control as well as being useful for pest control; wireworms, the larvae of the common click beetle, are very destructive pests of newly ploughed grassland, and repeated cultivation exposes them to the birds and other predators that feed on them.

Cultivation by ploughing exposes insect pests to predators such as black-headed gulls.

Crop rotation can help to control pests by depriving them of their host plants. It is a major tactic in the control of corn rootworm, and has reduced early season incidence of Colorado potato beetle by as much as 95%.

Trap Cropping

A trap crop is a crop of a plant that attracts pests, diverting them from nearby crops. Pests aggregated on the trap crop can be more easily controlled using pesticides or other methods. However, trap-cropping, on its own, has often failed to cost effectively reduce pest densities on large commercial scales, without the use of pesticides, possibly due to the pests' ability to disperse back into the main field.

Pesticides

Spraying pine logs with insecticideagainst Ips sexdentatus, a pine engraver beetle

Pesticides are applied to crops by agricultural aircraft, tractor-mounted crop sprayers or as seed dressings to control pests. However, successful control by pesticides is not easy; the right formulation must be chosen, the timing is often critical, the method of application is important, adequate coverage and retention on the crop are necessary. The killing of natural enemies of the target pest should be minimised. This is particularly important in countries where there are natural reservoirs of pests and their enemies in the countryside surrounding plantation crops, and these co-exist in a delicate balance. Often in less-developed countries, the crops are well adapted to the local situation and no pesticides are needed. Where progressive farmers are using fertilisers to grow improved crop varieties, these are often more susceptible to pest damage, but the indiscriminate application of pesticides may be detrimental in the longer term.

The efficacy of chemical pesticides tends to diminish over time. This is because any organism that manages to survive the initial application will pass on its genes to its offspring and a resistant strain will be developed. In this way, some of the most serious pests have developed resistance and are no longer killed by pesticides that used to kill their ancestors. This necessitates higher concentrations of chemical, more frequent applications and a movement to more expensive formulations.

Pesticides are formulated to kill pests, but many have detrimental effects on non-target species; of particular concern is the damage done to honey-bees, solitary bees and other pollinating insects and in this regard, the time of day when the spray is applied can be important. The widely used neonicotinoids have been banned on flowering crops in some countries because of their effects on bees. Some pesticides may cause cancer and other health problems in humans, as well as being harmful to wildlife. There can be acute effects immediately after exposure or chronic effects after continuous low-level, or occasional exposure. Maximum residue limits for pesticides in foodstuffs and animal feed are set by many nations.

Spruce budworm (adult and pupa shown), a serious pest of forests, can be monitored using pheromone traps.

Forestry

Forest pests present a significant problem because it is not easy to access the canopy and monitor pest populations. In addition, forestry pests such as bark beetles, kept under control by natural enemies in their native range, may be transported large distances in cut timber to places where they have no natural predators, enabling them to cause extensive economic damage. Pheromone traps have been used to monitor pest populations in the canopy. These release volatile chemicals that attract males. Pheromone traps can detect the arrival of pests or alert foresters to outbreaks. For example, the spruce budworm, a destructive pest of spruce and balsam fir, has been monitored using pheromone traps in Canadian forests for several decades. In some regions, such as New Brunswick, areas of forest are sprayed with pesticide to control the budworm population and prevent the damage caused during outbreaks.

Using organic pesticides is a highly efficient and effective way of controlling pests as it allows farmers to turn agricultural outputs into natural pesticides and does not affect their health or damage the crops. While it is common for most Indian farmers to use neem as a natural pesticide, farmers in Nepal spray 'Zhol Mol' (an organic liquid pesticide made of neem leaves), Timur (a Nepali spice), garlic, livestock urine, and water to their vegetables and fruits.

Benefits of Organic Pest Control

1) It's better for the environment

This is the core benefit of organic pest control. Whereas chemical sprays may be effective, they often have harmful effects on the surrounding environment (including animals, soil, and water when it becomes runoff), organic pest control products are typically much safer and more beneficial for the environment overall. Many organic sprays use food or plant-based oils as their active ingredients, giving you a natural way to keep pests off without damaging the surrounding environment.

2) You'll achieve healthier long-term results

Organic pesticides are typically effective for a longer period of time than chemical sprays. You've probably been through the routine with chemical pesticides before: apply once, apply again 30 days later, 30 days after that, etc.

The repeated applications can also damage your overall lawn or garden long-term and make it harder for any plants – not just weeds – to grow.

3) Pests can't resist it

It's been proven that chemical sprays and pesticides can lose their effectiveness over time. This simply comes down to nature and the passing on of beneficial traits: if certain insects in a species aren't killed off by the pesticides, only those with traits that make them resistant to it get passed on. Over time, the majority of the population may be resistant to that particular chemical.

Since organic treatments are biologically based rather than artificially created, pests are less likely to develop a resistance.

Mating Disruption

Many species of insects communicate by using a variety of chemicals. Chemical signals that elicit a response from other members of the same species are called "pheromones". "Sex pheromones" attract one sex to the other so that mating can take place and are relatively common in the insect order Lepidoptera (moths and butterflies). Sex pheromones are a complex mixture of chemicals and each species has its own specific blend. In most cases it is the female moth that emits the sex pheromones and the male that follows the pheromone trail (or plume) to find the female.

How does Mating Disruption Work?

Synthetically produced sex pheromones have been successfully used for decades to monitor insect activity patterns, and are extremely valuable tools in Integrated Pest Management (IPM) programs. Synthetically produced pheromones used in the manufacture of "lures" (for pheromone traps) are typically blends of major chemical components, along with some minor components, which attempt to mimic the effects of naturally produced pheromones. The more closely the blend matches that produced by a pheromone emitting female (known as a "calling" female), the greater the response by searching males.

Mating disruption (MD) technology uses synthetically produced chemicals in large amounts to confuse males and limit their ability to locate calling females; the blends, however, are often restricted to major components released by females (since the goal is to "disrupt" rather than "attract").

There are several mechanisms that may be at work in mating disruption, each of which may vary in importance as a function of the type of dispenser being used and the insect species. The release of sufficiently large quantities of synthetic sex pheromone into the crop atmosphere interferes with mate location by:

- masking the natural pheromone plumes
- affecting the males' ability to respond to calling females
- causing the male to follow "false pheromone trails" at the expense of finding mates.

By introducing many sources of the sex pheromone into the ecosystem, the probability of the male finding the female is reduced, as is the likelihood of successful mating. As a result, mating is either delayed (with a subsequent negative affect on overall fertility) or prevented. If female moths do not mate, they cannot lay fertile eggs and, if their mating is delayed, they will lay fewer fertilized eggs in their lifetime. Consequently, the subsequent population is reduced, and fewer larvae are present to cause crop damage.

Mating disruption offers a much different approach to insect pest management than traditional insecticides. Conventional programs typically use insecticides to target the damaging life stage (in most cases, the larva). In contrast, pheromones target the reproductive life stage (the adult), thus preventing the development of the damaging life stage. Pheromones used in mating disruption are species-specific and are thus highly selective. They are generally non-toxic and will not control other pests. It is important to thoroughly understand this fundamental difference before beginning a mating disruption program.

Use of Mating Disruption Products

Before starting to use any mating disruption products for pest management, discuss your particular situation with a qualified crop consultant experienced with MD. A site assessment to determine whether your field/orchard/ vineyard is suitable for MD is a valuable first step towards success with this type of pest control product. Plan ahead and have a season-long monitoring program personally designed before the season begins. Mating disruption can be one part of a planned IPM program agreed upon before the growing season begins. Integrated Pest Management is a decision-making process that uses all necessary techniques to suppress pests effectively, economically and in an environmentally sound manner. If you plan on monitoring your MD treated crop yourself ensure you:

- have the time for regular pest scouting
- know how to identify common insects and their damage, and
- know of any established thresholds for action if other pests become damaging.

Mating disruption works best if large areas are treated with pheromones. Ten acres is a good minimum size for a successful MD program but even larger areas are preferable. Cooperation among neighbouring farmers is probably the best way to approach an area-wide pest management program using MD. Square or rectangular blocks are best for using MD. Damage in MD blocks is typically associated with border areas (where mated females from untreated or poorly managed adjacent areas land and lay eggs). For this reason, long, narrow sites are not recommended. This technology cannot be "tried out" on a few rows or used in a small backyard planting - that is a sure path to failure with MD products.

Another important prerequisite for the successful use of MD is a low to moderate population level of the target pest. Mating disruption confuses males and makes it hard for them to find females. If the pest population in an MD site is very high to begin with, chance encounters and matings could still occur despite the use of sex pheromones. Therefore, this technology is frequently used in conjunction with insecticides. If possible, reduce populations of the pest through well timed and properly directed insecticide treatments, best management practices where appropriate, and rational cultural controls where feasible. This management can take place in years preceding the adoption of MD technology, or in addition to MD during the growing season. Once a pest population is consistently low through the combined use of cultural, insecticide and MD practices, it may be possible to manage the pest through MD and cultural controls alone.

There are a number of different types of hand-applied MD products that may be commercially available.

Hand-applied pheromone dispenser: twist-tie type.

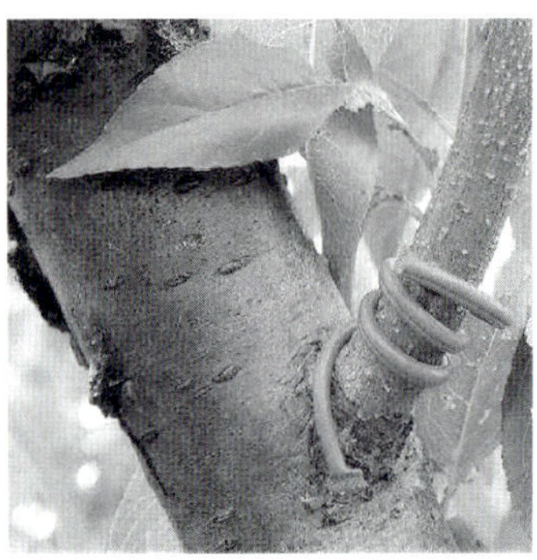

Twist-tie dispenser coiled on branch.

Hand-applied pheromone dispenser: clip type.

Hand-Applied Dispensers

Hand-applied dispensers may be available for certain fruit crop pests in Ontario in two main forms: most commonly used "twist ties" (sometimes called "tube-type") and "clips" (or "ampoules"). Different products may have very different rates of application but will always have a "per acre" or "per hectare" rate.

- Place dispensers in the crop individually (no bunches in one area to save time) and spread out as evenly as possible through the area to be protected. Some dispensers may indicate a different rate for block borders and some may suggest a different height for dispensers at block edges.
- Hand-applied dispensers are placed throughout the orchard/vineyard/field and may or may not need to be reapplied during the season, depending on the product's lifespan and seasonal flight activity of the target pest.
- Different dispensers may also need to be installed at a particular height in an orchard or vineyard; consult the product labels to be sure. Some moths such as codling moth (Cydia pomonella (L.)) mate primarily in the tops of orchard trees. For MD to work against such pests, dispensers need to be near the top of the trees. Other moths, such as oriental fruit moth (Grapholita molesta (Busck)) can be disrupted fairly easily with dispensers hung in orchards at about head height or slightly higher.
- Twist-tie type dispensers can be twisted onto branches in two ways; twisted over themselves or coiled around the branch (figures above). Girdling of the branch or trunk of small trees can occur occasionally if dispensers are twisted on too tightly. Always leave enough room for growth and twist the dispensers just enough to keep them from falling off. Girdling is an infrequent event but some growers may choose to remove twist ties from very young trees at the end of the season or in the spring before application of new ties. Applying the dispensers as a coil around the branch avoids the problem of girdling entirely.
- Apply clip type dispensers (figure above) onto sturdy branches of 1-1.5 cm diameter. Clip dispensers placed on branches smaller than 0.5 cm may spin in high winds and work themselves off the branch. Re-hang fallen dispensers. There is some anecdotal evidence that certain species of birds (particularly crows) may occasionally investigate clip dispensers in orchards and remove them from the trees.
- Most twist-tie dispensers will remain on the crop indefinitely and are difficult to distinguish from new dispensers. Do not paint dispensers or alter them in any way in order to tell them apart from older dispensers. Any alterations to dispensers can affect the rate at which pheromones are released from the dispensers and may make MD ineffective.
- Workers should wear disposable latex gloves when handling dispensers. Although insect pheromones have no effect on humans, a very small percentage

of people may develop a skin rash when handling large numbers of dispensers for long periods of time.

- Carpenter's aprons dedicated to use for each type of pheromone dispenser are useful when applying dispensers. Do not contaminate clothing by carrying bundles of dispensers in pockets.
- Unopened packages can be stored for use later in the year or the following year in a pesticide storage refrigerator or freezer. Opened packages can be kept for use during the season if reapplication is necessary but must be double-bagged and sealed tightly for short-term cold storage.

Advantages of Mating Disruption

Reducing pesticide use can have numerous benefits.

Resistance Management

One important benefit of using MD is in the field of resistance management. Repeated use of any one pesticide or any number of pesticides within a single chemical family can lead to populations of insect pests that are no longer controlled by those pesticides. Pesticides are valuable and important tools for modern agricultural production, so any methods to slow or eliminate the onset of resistance are beneficial. Rotation between different chemical groups or families is one effective method of preventing resistance from occurring. Decreasing the exposure of insect populations to pesticides is another method of reducing resistance. A mating disruption program can reduce the need for insecticide treatments for specific insect pests, thus reducing the pests' exposure to certain insecticides and possibly promoting the preponderance of pesticide-susceptible populations.

Reduced Impact on Human Health

Insect pheromones and MD products do not have any adverse effect on humans. Thus, the use of MD is beneficial in situations where frequent travel in the crop is necessary. Mating disruption products have no or very short re-entry intervals (time after application when it is safe to re-enter the field). This can be beneficial for scheduling labour or if public entry to the field/orchard/ vineyard is an issue. Pheromone products have very low or no days-to-harvest intervals (or preharvest intervals) and there are no concerns regarding pheromone residue on produce.

No Residues

In cases where hand-applied pheromone dispensers and overhead irrigation are used, irrigation scheduling is simplified because there is no concern with washing the product off the crop.

Long Term Control

Using MD over several years to manage a particular insect pest may help lower the local population of that pest. Reduced matings and delayed matings can lower the local population over time only if the pest is not too mobile. If the pest can travel extensively, locally lowered populations will likely be replenished from nearby sources.

No Non-Target Effects

Mating disruption products have a very good fit in IPM programs. They are species-specific and, for the most part, have no effect on other insect species. The compounds used are naturally occurring, generally non-toxic, and are safe to beneficials and other non-target organisms. Although the rise of secondary pests after the removal of broad-spectrum insecticides is a possibility, an equally likely scenario is that the reduction of insecticide use can lead to an increase in populations of beneficial insects, spiders and mites. An increase in beneficials can lead to further reductions in various insect pest populations in many cases.

Monitoring in MD Treated Crops

Monitoring or "pest scouting" is a key component of any IPM program. Regular, competent pest monitoring allows growers to base pest management decisions on actual pest pressure rather than on inaccurate and inefficient calendar spray schedules. Using independent crop consultants and monitoring services is highly recommended for all IPM programs.

Monitoring for many types of pests (especially moth pests) is often done using pheromone-baited sticky traps. These are excellent tools for monitoring population changes such as onset of first flight and changes in pest levels through the season. However, they are not particularly good at predicting ultimate crop damage levels. Because these traps use sex pheromones to attract male moths, they are not useful for monitoring moth populations in pheromone-treated (MD) sites.

- Pheromone-baited sticky traps should still be used in MD sites as "sentinel traps". If the MD product is emitting sufficient pheromone to prevent the location of calling females by searching males, few males should be caught in sentinel traps.
- Males are sometimes trapped in sentinel traps placed at MD block borders because the level of synthetic sex pheromone in the air at these locations is often lower than within the MD block and because males may fly in from adjacent unmanaged areas, landing first in border areas.
- Regular pest monitoring with sticky traps should not be done in blocks which have used MD the previous year; there may still be enough pheromone emitted

from previous year's dispensers to cause trap catch reductions but not necessarily enough to prevent moths from finding mates. For some pests, including codling moth, trapping can be accomplished in MD blocks using "super lures" with 10 times the amount of pheromone normally found in standard products. The information provided by these lures can be useful in timing any required supplemental insecticide sprays where area-wide MD programs are adopted.

Monitoring in MD blocks with sentinel traps is only one part of a comprehensive monitoring program. Thorough scouting for damage caused by the pest being managed with MD is also vital.

- Counts of indirect and direct damage may be required. Oriental fruit moth, for example, attacks primarily shoots in the first generation, moving onto fruit in subsequent generations. Shoot "strikes" caused by first generation larvae provide an early indication of hot spots and potential failures in MD sites, where thresholds are exceeded.
- Similarly, checks for damage by secondary pests are critical. These should be planned before the season begins and should be performed on a regular, rational schedule.
- Consultants should make growers aware of thresholds for damage for the pest targeted with MD and for other pests that may be present.
- In some cases, if early damage assessments show that the MD program is not providing sufficient protection for the crop, growers may be able to supplement the program with recommended insecticides to reduce economic loss.

Limitations of Mating Disruption

There are some limitations to MD programs, just as there are to any pest management options. The following limitations are listed in order of probable importance.

Size of Area Treated

The minimum suggested size for areas of MD is 2-4 ha (5-10 acres). Some product labels list 2 ha (5 acres) as the minimum size but experience with MD demonstration blocks in Niagara suggests that 4 ha (10 acres) are required to achieve success with MD. Neighbouring growers who cooperate and adopt wide area MD programs have a much better record of success with this technology. Mating disruption will not work on a few rows or on a few backyard trees.

Timing of Application

One of the most frequent causes for failure of MD products to suppress mating is incorrect timing of installing dispensers. As a general rule, MD products need to be on the crop before first flight of the pest species in the spring. Some programs have relied on the use of an insecticide directed at the first generation to reduce populations to a low level, with subsequent populations managed by MD (alone or with supplemental insecticides required as indicated by monitoring). There are some other restrictions and strategies depending on the particular pest and product being used so it is best to work closely with product representatives and crop consultants familiar with MD.

Shape of Area Treated

Mating disruption blocks should be as square as possible. Do not use this technology on long, narrow sites or a few rows of crop. Pheromones are volatile and move with airflow, so the pheromone product will not remain in sufficient quantities on narrow sites in normal wind conditions. In addition, individuals migrating from adjacent areas tend to land in border rows first. Long, narrow areas have a larger vulnerable border (relative to the total area treated) than do square blocks. Consequently, a larger portion of the crop remains at risk from migrating mated females.

Initial Pest Population Level

It is important for pest levels to be low to moderate for MD to work. High pest populations in an MD treated site can allow males and females to meet and mate by chance even in the presence of high levels of synthetic sex pheromone. In some cases, efforts must be made for several years to lower pest populations by using rotational strategies of registered insecticides and cultural controls. Pest levels are different across Ontario so consult with local crop consultants and other agricultural support personnel to devise the best strategy for lowering local populations before beginning an MD program.

Rates of Mating Disruption Products

Use mating disruption products, like any pest control products, at the recommended label rate. Research is being carried out worldwide on the lowest rates at which various MD products are effective and information on those projects is often available before final results are obtained.

Immigration of Mated Female Pests

Female pests may mate outside of MD blocks, enter the blocks and lay fertile eggs. How much this affects an MD block depends on many factors such as: how many mated females are in the vicinity of the MD block, how far they can fly, what time of year they are most active, how attractive the MD block is, how many alternate and wild hosts the pest has, and how close the MD block is to poorly managed or abandoned orchards. Immigration of gravid (mated) females may also increase late in the season when fewer "targets" (e.g.; ripe fruit) remain. Late season varieties may be particularly vulnerable to infestation through the migration of mated females.

Many pest insects will not travel very far in their lifetime and will not leave nearby blocks if there is sufficient suitable host plant material. Some however, travel considerable distances. Monitoring MD blocks is therefore critical to ensure damage is not occurring from the offspring of immigrating pests. For cross-commodity pests such as oriental fruit moth (stone and pome fruit), seasonal management strategies may vary between crops. Growers and crop consultants must consider when the crops are harvested relative to the seasonal activity of the pest.

Internal Sources of Pests

Cull piles and bin piles within an orchard/vineyard/field can also lead to infestations. Overwintering survival may be greater under these conditions. Males and females emerging in bin storage piles can easily find each other by chance in spite of pheromone treatment of the surrounding area. Sanitation involving the removal and disposal of infested fruit is key to reducing internal pest populations.

Secondary Pests

In some cases, insecticides targeted at major pests may control minor pests. Mating disruption programs for major pests should reduce insecticide use for control of major pests. Therefore, some minor pests may rise to damaging levels in MD blocks. It is also possible that reductions in insecticide use may allow beneficial insects and mites to thrive and provide natural biological control of minor pests.

The key to preventing economic damage from secondary pests is vigilant, regular, competent monitoring of MD areas. Monitoring provides early identification of damage from secondary pests and allows growers to take appropriate action before economic thresholds are reached. All IPM programs, including those using MD, need to be flexible to respond to the changing dynamics of pest complexes.

Size of Crop Plants

First year plantings of orchards are not recommended for MD. Pheromones emitted from dispensers will dissipate rapidly in situations where little foliage is present. However, a combination of insecticides and MD in young plantings may help lower local pest populations making MD more effective in subsequent years.

Temperature

Release of pheromones from fixed-point dispensers is temperature dependant. Below 10°C, little pheromone is emitted but that is not a problem as insects will be relatively inactive at low temperatures. High temperatures cause more rapid emission of pheromones from dispensers. The length of life of MD products may be shortened with prolonged high temperatures. In general, this should not be a problem in summers and only becomes a concern with consistent temperatures above 30°C-35°C.

Wind and Slope

Areas with consistent high winds are not suitable for MD because wind will move the pheromones away from the crop and the concentration of pheromones necessary for MD will not be maintained. Pheromones are heavier than air and can, in low wind situations, flow down steep slopes, leaving upper areas of a crop unprotected. Neither of these situations is typical for most areas in Ontario. One solution for these problems could be to increase the number of dispensers in upwind areas or on higher slopes.

Fly-killing Device

A fly-killing device is used for pest control of flying insects, such as houseflies, wasps, moths, gnats, and mosquitos. Most such devices are lethal to insects, but some designs (such as certain bug vacuums) can be used for live capture and later release.

Flyswatter

Fly Gun

A typical flyswatter

A flyswatter (or fly swatter) usually consists of a small rectangular or round sheet (some 4 inches (10 cm) across) of lightweight, flexible, vented material (usually thin metallic, rubber, or plastic mesh), attached to a lightweight wire or plastic handle about 1 foot (0.30 m) to 2 feet (0.61 m) long. The venting or perforations minimize the disruption of air currents, which can be detected by the fly and allow it to escape, and also reduce air resistance, making it easier to hit a fast-moving target such as a fly.

A flyswatter is ideally lightweight and stiff, allowing quick acceleration to overcome the fast reaction time of the fly, while also minimizing damage caused by hitting other objects. The flyswatter usually works by mechanically crushing the fly against a hard surface, after the user has waited for the fly to land somewhere. However, some skilled users can injure or stun an airborne insect in mid-flight by whipping the swatter through the air at an extreme speed.

Fly Bottle

Three fly bottles from Central Europe, beginning of the 20th century

A fly bottle or glass flytrap is a passive trap for flying insects. In the Far East, it is a large bottle of clear glass with a black metal top in which there is a hole. An odorous bait is placed in the bottom of the bottle in the form of pieces of meat. Flies enter the bottle in search of food and are then unable to escape because their phototaxis behavior leads them anywhere in the bottle except to the darker top where the entry hole is.

A European fly bottle is more conical, with small feet that raise it half an inch, with a trough about an inch wide and deep that runs inside the bottle all around the central opening at the bottom of the container. In use, the bottle is stood on a plate and some sugar is sprinkled on the plate to attract the flies, which eventually fly up into the bottle, whose trough is filled with beer or vinegar, into which the flies fall and drown. In the past, the trough was also sometimes filled with a dangerous mixture of milk, water, and arsenic or mercury chloride.

Variants of these bottles are the agricultural fly traps used to fight the Mediterranean fruit fly and the olive fly, which have been in use since the 1930s. They are smaller, without feet, and the glass is thicker for rough outdoor usage, often involving suspension in a tree or bush. Modern versions of this device are often made of plastic, and can be purchased in some hardware stores. They can also be improvised from disposable plastic drink bottles.

Disposable Fly Traps

Disposable fly traps are small "use and throw away" fly traps. The traps are disposable plastic bags containing some attractant, generally made of flavoring agents that are non-toxic. Water and direct sunlight are used to activate the attractant, which emits a smell to lure the flies. Insects enter the trap and drown in the water inside.

Glue Board

A glue board is a capture device with a strong adhesive. A small card covered in sticky adhesive is situated in an enclosure so that when the flies come into contact with it, they remain stuck and die there. A reusable glue board may be renewed through the use of vegetable oil, and then removing the oil with dishwashing detergent and a water rinse. Alternatively, the card is disposed of and completely replaced periodically.

Flypaper

Flypaper (also known as fly paper, fly sticker, fly strip, fly ribbon, or fly tape) attracts flies to adhesive so that they can be trapped. The exposed adhesive strip makes it more stick-prone than an enclosed glue board. To avoid accidental entanglement with humans, the strips are often hung in relatively inaccessible spaces, such as near ceilings. One type of fly strip is packaged in a small cardboard tube with a pin on the top. It is used by pulling the pin off the top (usually covered with wax), removing the adhesive "fly strip" and using the pin to attach it to a ceiling, with the tube dangling below as a small weight. Flypaper is not reused, but is replaced when it loses effectiveness.

A sticky fly ribbon

Flypaper is often impregnated with a slightly odorous chemical to attract more flies. The attractiveness of flypaper to other insects (such as mosquitos and biting midges) is sometimes enhanced by shining a small portable electric light on the sticky surface.

Bug Vacuum

A bug vacuum (bug vac or aspirator) is a type of small but powerful portable vacuum cleaner, usually with internal batteries. The motor starts quickly and generates strong suction, trapping the flying insect inside the device. The insect may be captured on an adhesive internal surface, or simply held inside the device until it dehydrates and dies.

Some bug vacuums feature non-lethal designs which keep trapped insects inside, but do not otherwise harm them, allowing their later release. These devices are popular with amateur and professional entomologists, and with persons who wish to avoid the killing of insects.

Fan-based Trap

This design uses a continuously-running electric fan to suck in flying insects (especially mosquitos and gnats, which are weak fliers), which are then trapped by a fine mesh grid or bag. Unable to escape the constant airflow, the insects quickly dehydrate and die. Some variant designs use carbon dioxide, ultraviolet light, or chemical scent to attract insects to the trap. Other designs rely on the natural carbon dioxide or scents emitted by people, pets, or livestock to attract pests, and simply collect flying insects as they wander close enough to be sucked in. In addition, the continuous breeze produced by a common electric fan has been found to discourage mosquitos from landing and biting, even without trapping or killing the insects.

Bug Zapper

A bug zapper electric grid (fly zapper) kills insects by electrocution from high voltage on an electric grid. Bug zappers are generally small appliances intended for use in a fixed location, as distinguished from hand held electric flyswatters.

Electric Flyswatter

An electric flyswatter

Closeup view of three layer grid: mesh and rods oppositely charged

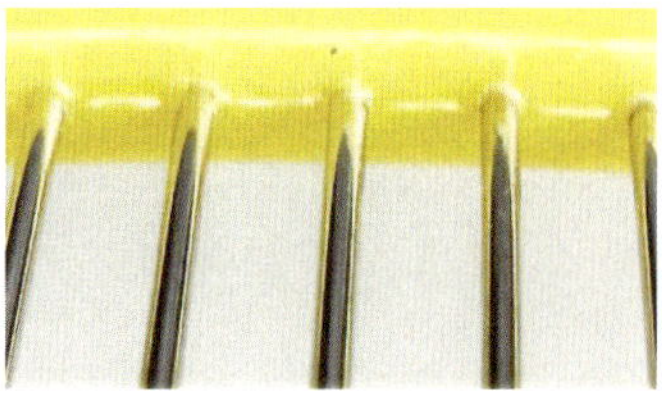

Closeup view of single layer grid: odd and even rods oppositely charged

Electric flyswatters (sometimes called mosquito bats, racket zappers, or zap rackets) are hand-held devices that resemble badminton rackets or tennis rackets, which became popular worldwide in the late 1990s. US Patent 5,519,963 was awarded to Taiwanese inventor Tsao-i Shih in 1996 for such a device. The handle contains a battery-powered high-voltage generator. The circuit is composed of an electronic oscillator, a step-up transformer and a voltage multiplier, similar to the circuit in an electroshock weapon or stun gun, but with much lower power.

The grid of the flyswatter is electrically charged to a voltage of between 500 and 1500 volts (V) of direct current, activated by pressing and holding a button. When the electrically conductive body of a fly nearly bridges the gap between electrodes, a spark jumps through the fly. A capacitor attached to the electrodes discharges during the spark, and this initial discharge usually stuns or kills the fly. If the button is still pressed, the continuous current will roast and kill the fly. Many flyswatters have a three-layer grid to prevent people from touching both electrodes. The outermost grids or rods are at the same electrical potential, and are open enough to allow an insect to contact the inner charged grid.

Most electric flyswatters conform to electrical safety standards for humans:

- a limit on the charge stored in the capacitor: a discharge of less than 45 microcoulombs (μC) is considered safe, even in the unlikely scenario that the current from a flyswatter would be flowing from one arm to the other arm, partly through the heart. This means that the capacitor of a 1000 V flyswatter should be less than 45 nanofarads (nF). Due to this precaution for humans, the initial shock is usually inadequate to kill flies, but will stun them for long enough that they can be disposed of.
- A limit on the current after the initial discharge: the maximal continuous current of most flyswatters is less than 5 milliamps (mA). This current is safe, even when flowing from one arm to the other arm of a human.

An advantage over conventional flyswatters is that the electrical models do not have to mechanically crush the fly against a hard surface to kill it, avoiding the smeared mess this can create. Also, the electrical grid can be relatively open, reducing air resistance and a rush of air that often deflects smaller insects around conventional swatters. Because of this, electric swatters can also be very effective in killing airborne mosquitos.

Biological Pest Control

Biological control, biocontrol, or biological pest control is a method of suppressing or controlling the population of undesirable insects, other animals, or plants by the introduction, encouragement, or artificial increase of their natural enemies to economically non–important levels. It is an important component of integrated pest management (IPM) programs (Weeden et al. 2007).

The biological control of pests and weeds relies on predation, parasitism, herbivory, or other natural mechanisms. Therefore, it is the active manipulation of natural phenomena in serving human purpose, working harmoniously with nature. A successful story of biological control of pests refer to the human beings' capability to depict natural processes for their use and can be the most harmless, non–polluting, and self–perpetuating control method.

Overview

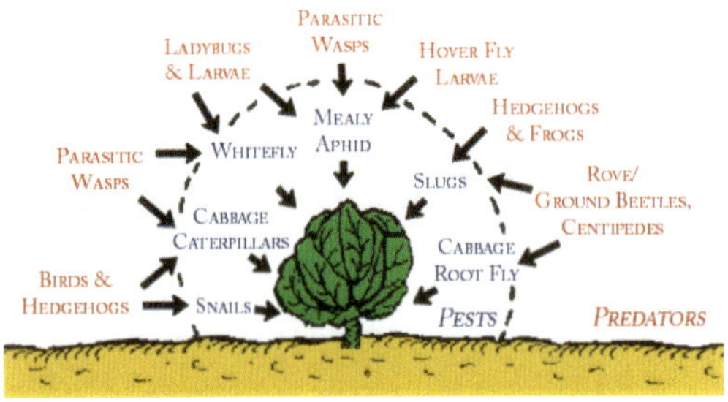

Natural enemies of cabbage pests

Predatory Polistes wasp looking for bollworms or other caterpillars on a cotton plant

In biological control, the reduction of pest populations is achieved by actively using natural enemies.

Natural enemies of the pests, also known as biological control agents, include predatory and parasitoidal insects, predatory vertebrates, nematode parasites, protozoan parasites, and fungal, bacterial, as well as viral pathogens (Metcalf et al. 1973). Biological control agents of plant diseases are most often referred to as antagonists. Biological control agents of weeds include herbivores and plant pathogens. Predators, such as lady beetles and lacewings, are mainly free–living species that consume a large number of prey during their lifetime. Parasitoids are species whose immature stage develops on or within a single insect host, ultimately killing the host. Most have a very narrow host range. Many species of wasps and some flies are parasitoids. Pathogens are disease–causing organisms including bacteria, fungi, and viruses. They kill or debilitate their host and are relatively specific to certain pest or weed groups.

Types of Biological Pest Control

There are three basic biological pest control strategies: importation (classical biological control), augmentation and conservation.

Importation

Rodolia cardinalis, the vedalia beetle, was imported from Australia to California in the 19th century, successfully controlling cottony cushion scale.

Importation or classical biological control involves the introduction of a pest's natural enemies to a new locale where they do not occur naturally. Early instances were often unofficial and not based on research, and some introduced species became serious pests themselves.

To be most effective at controlling a pest, a biological control agent requires a colonizing ability which allows it to keep pace with changes to the habitat in space and time. Control is greatest if the agent has temporal persistence, so that it can maintain its population even in the temporary absence of the target species, and if it is an opportunistic forager, enabling it to rapidly exploit a pest population.

Joseph Needham noted a Chinese text dating from 304 AD, *Records of the Plants and Trees of the Southern Regions*, by Hsi Han, which describes mandarin oranges protected by large reddish-yellow citrus ants which attack and kill insect pests of the orange trees. The citrus ant (*Oecophylla smaragdina*) was rediscovered in the 20th century, and since 1958 has been used in China to protect orange groves.

One of the earliest successes in the west was in controlling *Icerya purchasi* (cottony cushion scale) in Australia, using a predatory insect *Rodolia cardinalis* (the vedalia beetle). This success was repeated in California using the beetle and a parasitoidal fly, *Cryptochaetum iceryae*.

Damage from *Hypera postica*, the alfalfa weevil, a serious introduced pest of forage, was substantially reduced by the introduction of natural enemies. 20 years after their introduction the population of weevils in the alfalfa area treated for alfalfa weevil in the Northeastern United States remained 75 percent down.

Alligator weed was introduced to the United States from South America. It takes root in shallow water, interfering with navigation, irrigation, and flood control. The alligator weed flea beetle and two other biological controls were released in Florida, greatly reducing the amount of land covered by the plant. Another aquatic weed, the giant salvinia (*Salvinia molesta*) is a serious pest, covering waterways, reducing water flow and harming native species. Control with the salvinia weevil (*Cyrtobagous salviniae*) and the salvinia stem-borer moth (*Samea multiplicalis)* is effective in warm climates, and in Zimbabwe, a 99% control of the weed was obtained over a two-year period.

The invasive species *Alternanthera philoxeroides* (alligator weed) was controlled in Florida (U.S.) by introducing alligator weed flea beetle.

Small commercially reared parasitoidal wasps, *Trichogramma ostriniae*, provide limited and erratic control of the European corn borer (*Ostrinia nubilalis*), a serious pest. Careful formulations of the bacterium *Bacillus thuringiensis* are more effective.

The population of *Levuana iridescens*, the Levuana moth, a serious coconut pest in Fiji, was brought under control by a classical biological control program in the 1920s.

Augmentation

Hippodamia convergens, the convergent lady beetle, is commonly sold for biological control of aphids.

Augmentation involves the supplemental release of natural enemies that occur in a particular area, boosting the naturally occurring populations there. In inoculative release, small numbers of the control agents are released at intervals to allow them to reproduce, in the hope of setting up longer-term control, and thus keeping the pest down to a low level, constituting prevention rather than cure. In inundative release, in contrast, large numbers are released in the hope of rapidly reducing a damaging pest population, correcting a problem that has already arisen. Augmentation can be effective, but is not guaranteed to work, and depends on the precise details of the interactions between each pest and control agent.

An example of inoculative release occurs in the horticultural production of several crops in greenhouses. Periodic releases of the parasitoidal wasp, *Encarsia formosa*, are used to control greenhouse whitefly, while the predatory mite *Phytoseiulus persimilis* is used for control of the two-spotted spider mite.

The egg parasite *Trichogramma* is frequently released inundatively to control harmful moths. Similarly, *Bacillus thuringiensis* and other microbial insecticides are used in large enough quantities for a rapid effect. Recommended release rates for *Trichogramma* in vegetable or field crops range from 5,000 to 200,000 per acre (1 to 50 per square metre) per week according to the level of pest infestation. Similarly, nematodes that kill insects (that are entomopathogenic) are released at rates of millions and even billions per acre for control of certain soil-dwelling insect pests.

Conservation

The conservation of existing natural enemies in an environment is the third method of biological pest control. Natural enemies are already adapted to the habitat and to the target pest, and their conservation can be simple and cost-effective, as when nectar-producing crop plants are grown in the borders of rice fields. These provide nectar to support parasitoids and predators of planthopper pests and have been demonstrated to be so effective (reducing pest densities by 10- or even 100-fold) that farmers sprayed 70% less insecticides and enjoyed yields boosted by 5%. Predators of aphids were similarly found to be present in tussock grasses by field boundary hedges in England, but they spread too slowly to reach the centres of fields. Control was improved by planting a metre-wide strip of tussock grasses in field centres, enabling aphid predators to overwinter there.

An inverted flowerpot filled with straw to attract earwigs

Cropping systems can be modified to favor natural enemies, a practice sometimes referred to as habitat manipulation. Providing a suitable habitat, such as a shelterbelt, hedgerow, or beetle bank where beneficial insects such as parasitoidal wasps can live and reproduce, can help ensure the survival of populations of natural enemies. Things as simple as leaving a layer of fallen leaves or mulch in place provides a suitable food source for worms and provides a shelter for insects, in turn being a food source for such beneficial mammals as hedgehogs and shrews. Compost piles and stacks of wood can provide shelter for invertebrates and small mammals. Long grass and ponds support amphibians. Not removing dead annuals and non-hardy plants in the autumn allows insects to make use of their hollow stems during winter. In California, prune trees are sometimes planted in grape vineyards to provide an improved overwintering habitat or refuge for a key grape pest parasitoid. The providing of artificial shelters in the form of wooden caskets, boxes or flowerpots is also sometimes undertaken, particularly in gardens, to make a cropped area more attractive to natural enemies. For example, earwigs are natural predators which can be encouraged in gardens by hanging upside-down flowerpots filled with straw or wood wool. Green lacewings can be encouraged by using plastic bottles with an open bottom and a roll of cardboard inside. Birdhouses enable insectivorous birds to nest; the most useful birds can be attracted by choosing an opening just large enough for the desired species.

In cotton production, the replacement of broad-spectrum insecticides with selective control measures such as Bt cotton can create a more favorable environment for natural enemies of cotton pests due to reduced insecticide exposure risk. Such predators or parasitoids can control pests not affected by the Bt protein. Reduced prey quality and abundance associated increased control from Bt cotton can also indirectly decrease natural enemy populations in some cases, but the percentage of pests eaten or parasitized in Bt and non-Bt cotton are often similar.

Biological Control Agents

Predators

Lacewings are available from biocontrol dealers.

Predators are mainly free-living species that directly consume a large number of prey during their whole lifetime. Given that many major crop pests are insects, many of the predators used in biological control are insectivorous species. Lady beetles, and in particular their larvae which are active between May and July in the northern hemisphere, are voracious predators of aphids, and also consume mites, scale insects and small caterpillars. The spotted lady beetle (*Coleomegilla maculata*) is also able to feed on the eggs and larvae of the Colorado potato beetle (*Leptinotarsa decemlineata*).

The larvae of many hoverfly species principally feed upon aphids, one larva devouring up to 400 in its lifetime. Their effectiveness in commercial crops has not been studied.

Several species of entomopathogenic nematode are important predators of insect and other invertebrate pests. *Phasmarhabditis hermaphrodita* is a microscopic nematode that kills slugs. Its complex life cycle includes a free-living, infective stage in the soil where it becomes associated with a pathogenic bacteria such as *Moraxella osloensis*. The nematode enters the slug through the posterior mantle region, thereafter feeding and reproducing inside, but it is the bacteria that kill the slug. The nematode is available commercially in Europe and is applied by watering onto moist soil.

Predatory *Polistes* wasp searching for bollworms or other caterpillars on a cotton plant

Species used to control spider mites include the predatory mites *Phytoseiulus persimilis*, *Neoseilus californicus,* and *Amblyseius cucumeris*, the predatory midge *Feltiella acarisuga*, and a ladybird *Stethorus punctillum*. The bug *Orius insidiosus* has been successfully used against the two-spotted spider mite and the western flower thrips (*Frankliniella occidentalis*).

Predators including *Cactoblastis cactorum* (mentioned above) can also be used to destroy invasive plant species. As another example, the poison hemlock moth (*Agonopterix alstroemeriana)* can be used to control poison hemlock (*Conium maculatum*). During its larval stage, the moth strictly consumes its host plant, poison hemlock, and can exist at hundreds of larvae per individual host plant, destroying large swathes of the hemlock.

The parasitoid wasp *Aleiodes indiscretus* parasitizing a gypsy moth caterpillar, a serious pest of forestry

For rodent pests, cats are effective biological control when used in conjunction with reduction of "harborage"/hiding locations. While cats are effective at preventing rodent "population explosions", they are not effective for eliminating pre-existing severe infestations. Barn owls are also sometimes used as biological rodent control. Although there are no quantitative studies of the effectiveness of barn owls for this purpose, they are known rodent predators that can be used in addition to or instead of cats; they can be encouraged into an area with nest boxes.

Parasitoids

Parasitoids lay their eggs on or in the body of an insect host, which is then used as a food for developing larvae. The host is ultimately killed. Most insect parasitoids are wasps or flies, and many have a very narrow host range. The most important groups are the ichneumonid wasps, which mainly use caterpillars as hosts; braconid wasps, which attack caterpillars and a wide range of other insects including aphids; chalcid wasps, which parasitize eggs and larvae of many insect species; and tachinid flies, which parasitize a wide range of insects including caterpillars, beetle adults and larvae, and true bugs. Parasitoids are most effective at reducing pest populations when their host organisms have limited refuges to hide from them.

Encarsia formosa, widely used in greenhouse horticulture, was one of the first biological control agents developed.

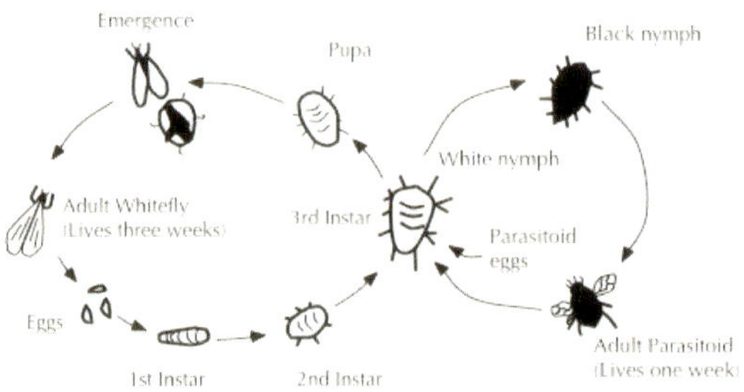

Life cycles of greenhouse whitefly and its parasitoid wasp *Encarsia formosa*

Encarsia formosa is a small predatory chalcid wasp which is a parasitoid of whitefly, a sap-feeding insect which can cause wilting and black sooty moulds in glasshouse vegetable and ornamental crops. It is most effective when dealing with low level infestations, giving protection over a long period of time. The wasp lays its eggs in young whitefly 'scales', turning them black as the parasite larvae pupate. *Gonatocerus ashmeadi* (Hymenoptera: Mymaridae) has been introduced to control the glassy-winged sharpshooter *Homalodisca vitripennis* (Hemiptera: Cicadellidae) in French Polynesia and has successfully controlled ~95% of the pest density.

Parasitoids are among the most widely used biological control agents. Commercially, there are two types of rearing systems: short-term daily output with high production of parasitoids per day, and long-term, low daily output systems. In most instances, production will need to be matched with the appropriate release dates when susceptible host species at a suitable phase of development will be available. Larger production facilities produce on a yearlong basis, whereas some facilities produce only seasonally. Rearing facilities are usually a significant distance from where the agents are to be used in the field, and transporting the parasitoids from the point of production to the point of use can pose problems. Shipping conditions can be too hot, and even vibrations from planes or trucks can adversely affect parasitoids.

The eastern spruce budworm is an example of a destructive insect in fir and spruce forests. Birds are a natural form of biological control, but the *Trichogramma minutum*, a species of parasitic wasp, has been investigated as an alternative to more controversial chemical controls.

Pathogens

Pathogenic micro-organisms include bacteria, fungi, and viruses. They kill or debilitate their host and are relatively host-specific. Various microbial insect diseases occur naturally, but may also be used as biological pesticides. When naturally occurring, these outbreaks are density-dependent in that they generally only occur as insect populations become denser.

Bacteria

Bacteria used for biological control infect insects via their digestive tracts, so they offer only limited options for controlling insects with sucking mouth parts such as aphids and scale insects. *Bacillus thuringiensis* is the most widely applied species of bacteria used for biological control, with at least four sub-species used against Lepidopteran (moth, butterfly), Coleopteran (beetle) and Dipteran (true fly) insect pests. The bacterium is available to organic farmers in sachets of dried spores which are mixed with water and sprayed onto vulnerable plants such as brassicas and fruit trees. Genes from *B. thuringiensis* have also been incorporated into transgenic crops, making the plants express some of the bacterium's toxins, which are proteins. These confer resistance to insect pests and thus reduce the necessity for pesticide use. If pests develop resistance to the toxins in these crops, *B. thuringiensis* will become useless in organic farming also. The bacterium *Paenibacillus popilliae* which causes milky spore disease has been found useful in the control of Japanese beetle, killing the larvae. It is very specific to its host species and is harmless to vertebrates and other invertebrates.

Fungi

Entomopathogenic fungi, which cause disease in insects, include at least 14 species that attack aphids. *Beauveria bassiana* is mass-produced and used to manage a wide variety of insect pests including whiteflies, thrips, aphids and weevils. *Lecanicillium* spp. are deployed against white flies, thrips and aphids. *Metarhizium* spp. are used against pests including beetles, locusts and other grasshoppers, Hemiptera, and spider mites. *Paecilomyces fumosoroseus* is effective against white flies, thrips and aphids; *Purpureocillium lilacinus* is used against root-knot nematodes, and 89 *Trichoderma* species against certain plant pathogens. *Trichoderma viride* has been used against Dutch elm disease, and has shown some effect in suppressing silver leaf, a disease of stone fruits caused by the pathogenic fungus *Chondrostereum purpureum*.

Green peach aphid, a pest in its own right and a vector of plant viruses, killed by the fungus *Pandora neoaphidis* (Zygomycota: Entomophthorales) Scale bar = 0.3 mm.

The fungi *Cordyceps* and *Metacordyceps* are deployed against a wide spectrum of arthropods. *Entomophaga* is effective against pests such as the green peach aphid.

Several members of Chytridiomycota and Blastocladiomycota have been explored as agents of biological control. From Chytridiomycota, *Synchytrium solstitiale* is being considered as a control agent of the yellow star thistle (*Centaurea solstitialis*) in the United States.

Viruses

Baculoviruses are specific to individual insect host species and have been shown to be useful in biological pest control. For example, the Lymantria dispar multicapsid nuclear polyhedrosis virus has been used to spray large areas of forest in North America where larvae of the gypsy moth are causing serious defoliation. The moth larvae are killed by the virus they have eaten and die, the disintegrating cadavers leaving virus particles on the foliage to infect other larvae.

A mammalian virus, the rabbit haemorrhagic disease virus was introduced to Australia to attempt to control the European rabbit populations there. It escaped from quarantine and spread across the country, killing large numbers of rabbits. Very young animals survived, passing immunity to their offspring in due course and eventually producing a virus-resistant population. Introduction into New Zealand in the 1990s was similarly successful at first, but a decade later, immunity had developed and populations had returned to pre-RHD levels.

Oomycota

Lagenidium giganteum is a water-borne mould that parasitizes the larval stage of mosquitoes. When applied to water, the motile spores avoid unsuitable host species and search out suitable mosquito larval hosts. This mould has the advantages of a dormant phase, resistant to desiccation, with slow-release characteristics over several years. Unfortunately, it is susceptible to many chemicals used in mosquito abatement programmes.

Competitors

The legume vine *Mucuna pruriens* is used in the countries of Benin and Vietnam as a biological control for problematic *Imperata cylindrica* grass: the vine is extremely vigorous and suppresses neighbouring plants by out-competing them for space and light. *Mucuna pruriens* is said not to be invasive outside its cultivated area. *Desmodium uncinatum* can be used in push-pull farming to stop the parasitic plant, witchweed (*Striga*).

The Australian bush fly, *Musca vetustissima*, is a major nuisance pest in Australia, but native decomposers found in Australia are not adapted to feeding on cow dung, which is where bush flies breed. Therefore, the Australian Dung Beetle Project (1965–1985), led by George Bornemissza of the Commonwealth Scientific and Industrial Research Organisation, released forty-nine species of dung beetle, to reduce the amount of dung and therefore also the potential breeding sites of the fly.

Combined use of Parasitoids and Pathogens

In cases of massive and severe infection of invasive pests, techniques of pest control are often used in combination. An example is the emerald ash borer, *Agrilus planipennis*, an invasive beetle from China, which has destroyed tens of millions of ash trees in its introduced range in North America. As part of the campaign against it, from 2003 American scientists and the Chinese Academy of Forestry searched for its natural enemies in the wild, leading to the discovery of several parasitoid wasps, namely *Tetrastichus planipennisi*, a gregarious larval endoparasitoid, *Oobius agrili*, a solitary, parthenogenic egg parasitoid, and *Spathius agrili*, a gregarious larval ectoparasitoid. These have been introduced and released into the United States of America as a possible biological control of the emerald ash borer. Initial results for *Tetrastichus planipennisi* have shown promise, and it is now being released along with *Beauveria bassiana*, a fungal pathogen with known insecticidal properties.

Difficulties

Many of the most important pests are exotic, invasive species that severely impact agriculture, horticulture, forestry and urban environments. They tend to arrive without their co-evolved parasites, pathogens and predators, and by escaping from these, populations may soar. Importing the natural enemies of these pests may seem a logical move but this may have unintended consequences; regulations may be ineffective and there may be unanticipated effects on biodiversity, and the adoption of the techniques may prove challenging because of a lack of knowledge among farmers and growers.

Side Effects

Biological control can affect biodiversity through predation, parasitism, pathogenicity, competition, or other attacks on non-target species. An introduced control does not always target only the intended pest species; it can also target native species. In Hawaii during the 1940s parasitic wasps were introduced to control a lepidopteran pest and the wasps are still found there today. This may have a negative impact on the native ecosystem; however, host range and impacts need to be studied before declaring their impact on the environment.

Vertebrate animals tend to be generalist feeders, and seldom make good biological control agents; many of the classic cases of "biocontrol gone awry" involve vertebrates. For example, the cane toad (*Rhinella marina*) was intentionally introduced to Australia to control the greyback cane beetle (*Dermolepida albohirtum*), and other pests of sugar cane. 102 toads were obtained from Hawaii and bred in captivity to increase their numbers until they were released into the sugar cane fields of the tropic north in 1935. It was later discovered that the toads could not jump very high and so were unable to eat the cane beetles which stayed on the upper stalks of the cane plants. However, the toad thrived by feeding on other insects and soon spread very rapidly; it took over native amphibian habitat and brought foreign disease to native toads and frogs, dramatically reducing their populations. Also, when it is threatened or handled, the cane toad releases poison from parotoid glands on its shoulders; native Australian species such as goannas, tiger snakes, dingos and northern quolls that attempted to eat the toad were harmed or killed. However, there has been some recent evidence that native predators are adapting, both physiologically and through changing their behaviour, so in the long run, their populations may recover.

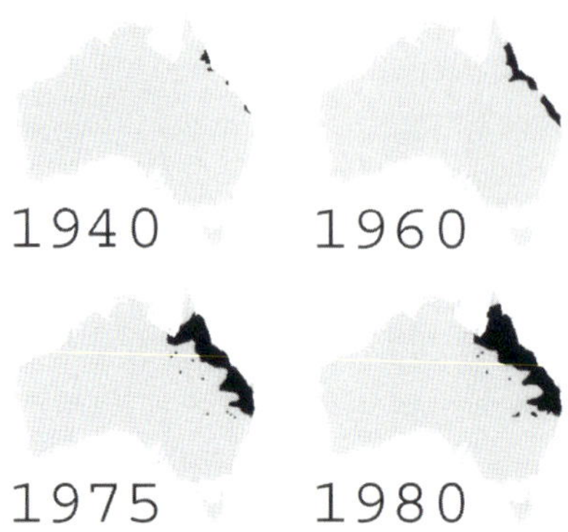

Cane toad (introduced into Australia 1935) spread from 1940 to 1980:
it was ineffective as a control agent. Its distribution has continued to widen since 1980.

Rhinocyllus conicus, a seed-feeding weevil, was introduced to North America to control exotic musk thistle (*Carduus nutans*) and Canadian thistle (*Cirsium arvense*). However, the weevil also attacks native thistles, harming such species as the endemic Platte thistle (*Cirsium neomexicanum*) by selecting larger plants (which reduced the gene pool), reducing seed production and ultimately threatening the species' survival.

The small Asian mongoose (*Herpestus javanicus*) was introduced to Hawaii in order to control the rat population. However, the mongoose was diurnal, and the rats emerged at night; the mongoose therefore preyed on the endemic birds of Hawaii, especially their eggs, more often than it ate the rats, and now both rats and mongooses threaten the birds. This introduction was undertaken without understanding the consequences of such an action. No regulations existed at the time, and more careful evaluation should prevent such releases now.

The sturdy and prolific eastern mosquitofish (*Gambusia holbrooki*) is a native of the southeastern United States and was introduced around the world in the 1930s and '40s to feed on mosquito larvae and thus combat malaria. However, it has thrived at the expense of local species, causing a decline of endemic fish and frogs through competition for food resources, as well as through eating their eggs and larvae. In Australia, control of the mosquitofish is the subject of discussion; in 1989 it was said that "biological population control is well beyond present capabilities", and this remains the position.

Grower Education

A potential obstacle to the adoption of biological pest control measures is growers sticking to the familiar use of pesticides. However, pesticides have a variety of undesired effects, including the development of resistance among pests, and the destruction of natural enemies; these may in turn enable outbreaks of pests of other species than the ones originally targeted, and on crops at a distance from those treated with pesticides. One method of increasing grower adoption of biocontrol methods involves letting them learn by doing, for example showing them simple field experiments, enabling them to observe the live predation of pests, or demonstrations of parasitised pests. In the Philippines, early season sprays against leaf folder caterpillars were common practice, but growers were asked to follow a 'rule of thumb' of not spraying against leaf folders for the first 30 days after transplanting; participation in this resulted in a reduction of insecticide use by 1/3 and a change in grower perception of insecticide use.

Plants to Regulate Insect Pests

Choosing a diverse range of plants for the garden can help to regulate pests in a variety of ways, including;

- Masking the crop plants from pests, depending on the proximity of the companion or intercrop.
- Producing olfactory inhibitors, odors that confuse and deter pests.
- Acting as trap plants by providing an alluring food that entices pests away from crops.
- Serving as nursery plants, providing breeding grounds for beneficial insects.
- Providing an alternative habitat, usually in a form of a shelterbelt, hedgerow, or beetle bank, where beneficial insects can live and reproduce. Nectar–rich plants that bloom for long periods are especially good, as many beneficials are nectivorous during the adult stage, but parasitic or predatory as larvae. A good example of this is the soldier beetle, which is frequently found on flowers as an adult, but whose larvae eat aphids, caterpillars, grasshopper eggs, and other beetles.

The following are plants often used in vegetable gardens to deter insects:

Plant	**Pests**
Basil	Repels flies and mosquitoes.
Catnip	Deters flea beetle.
Garlic	Deters Japanese beetle.
Horseradish	Deters potato bugs.
Marigold	The workhorse of pest deterrents. Discourages Mexican bean beetles, nematodes and others.
Mint	Deters white cabbage moth, ants.
Nasturtium	Deters aphids, squash bugs and striped pumpkin beetles.
Pot Marigold	Deters asparagus beetles, tomato worm, and general garden pests.
Peppermint	Repels the white cabbage butterfly.
Rosemary	Deters cabbage moth, bean beetles and carrot fly.
Sage	Deters cabbage moth and carrot fly.
Southernwood	Deters cabbage moth.
Summer Savory	Deters bean beetles.
Tansy	Deters flying insects, Japanese beetles, striped cucumber beetles, squash bugs and ants.
Thyme	Deters cabbage worm.
Wormwood	Deters animals from garden.

Pathogens to be used as Biopesticides

Various bacterial species are widely used in controlling the pests as well as weeds.

The best–known bacterial biological control which can be introduced in order to control butterfly caterpillars is *Bacillus thuringiensis,* popularly called *Bt*. This is available in sachets of dried spores, which are mixed with water and sprayed onto vulnerable plants such as brassicas and fruit trees. After ingestion of the bacterial preparation, the endotoxin liberated and activated in the midgut will kill the caterpillars, but leave other insects unharmed. There are strains of *Bt* that are effective against other insect larvae. *Bt. israelensis* is effective against mosquito larvae and some midges.

Viruses most frequently considered for the control of insects (usually sawflies and Lepidoptera) are the occluded viruses, namely NPV, cytoplasmic polyhedrosis (CPV), granulosis (GV), and entomopox viruses (EPN). They do not infect vertebrates, non–arthropod invertebrates, microorganisms, and plants. The commercial use of virus insecticides has been limited by their high specificity and slow action.

Fungi are pathogenic agents to various organisms including the pests and the weeds. This feature is intensively used in biocontrol. The entomopathogenic fungi, like *Metarhizium anisopliae, Beauveria bassiana,* and so forth cause death to the host by the secretion of toxins. A biological control being developed for use in the treatment of plant disease is the fungus *Trichoderma viride*. This has been used against Dutch Elm disease, and to treat the spread of fungal and bacterial growth on tree wounds. It may also have potential as a means of combating silver leaf disease.

Integrated Pest Management

Integrated pest management, or IPM, is a process you can use to solve pest problems while minimizing risks to people and the environment. IPM can be used to manage all kinds of pests anywhere—in urban, agricultural, and wildland or natural areas.

IPM is an ecosystem-based strategy that focuses on long-term prevention of pests or their damage through a combination of techniques such as biological control, habitat manipulation, modification of cultural practices, and use of resistant varieties. Pesticides are used only after monitoring indicates they are needed according to established guidelines, and treatments are made with the goal of removing only the target organism. Pest control materials are selected and applied in a manner that minimizes risks to human health, beneficial and nontarget organisms, and the environment.

How does IPM work

IPM focuses on long-term prevention of pests or their damage by managing the ecosystem.

With IPM, you take actions to keep pests from becoming a problem, such as by growing a healthy crop that can withstand pest attacks, using disease-resistant plants, or caulking cracks to keep insects or rodents from entering a building.

Rather than simply eliminating the pests you see right now, using IPM means you'll look at environmental factors that affect the pest and its ability to thrive. Armed with this information, you can create conditions that are unfavorable for the pest.

In IPM, monitoring and correct pest identification help you decide whether management is needed.

Monitoring means checking your field, landscape, forest, or building—or other site—to identify which pests are present, how many there are, or what damage they've caused. Correctly identifying the pest is key to knowing whether a pest is likely to become a problem and determining the best management strategy.

After monitoring and considering information about the pest, its biology, and environmental factors, you can decide whether the pest can be tolerated or whether it is a problem that warrants control. If control is needed, this information also helps you select the most effective management methods and the best time to use them.

IPM programs combine management approaches for greater effectiveness.

The most effective, long-term way to manage pests is by using a combination of methods that work better together than separately. Approaches for managing pests are often grouped in the following categories.

Biological Control

Biological control is the use of *natural enemies*—predators, parasites, pathogens, and competitors—to control pests and their damage. Invertebrates, plant pathogens, nematodes, weeds, and vertebrates have many natural enemies.

Cultural Controls

Cultural controls are practices that reduce pest establishment, reproduction, dispersal, and survival. For example, changing irrigation practices can reduce pest problems, since too much water can increase root disease and weeds.

Mechanical and Physical Controls

Mechanical and physical controls kill a pest directly, block pests out, or make the environment unsuitable for it. Traps for rodents are examples of mechanical control. Physical controls include mulches for weed management, steam sterilization of the soil for disease management, or barriers such as screens to keep birds or insects out.

Chemical Control

Chemical control is the use of pesticides. In IPM, pesticides are used only when needed and in combination with other approaches for more effective, long-term control. Pesticides are selected and applied in a way that minimizes their possible harm to people, nontarget organisms, and the environment. With IPM you'll use the most selective pesticide that will do the job and be the safest for other organisms and for air, soil, and water quality; use pesticides in bait stations rather than sprays; or spot-spray a few weeds instead of an entire area.

Applications

IPM is used in agriculture, horticulture, forestry, human habitations, preventive conservation and general pest control, including structural pest management, turf pest management and ornamental pest management.

Principles

An American IPM system is designed around six basic components:

- Acceptable pest levels—The emphasis is on *control*, not *eradication*. IPM holds that wiping out an entire pest population is often impossible, and the attempt can be expensive and unsafe. IPM programmes first work to establish acceptable pest levels, called action thresholds, and apply controls if those thresholds are crossed. These thresholds are pest and site specific, meaning that it may be acceptable at one site to have a weed such as white clover, but not at another site. Allowing a pest population to survive at a reasonable threshold reduces selection pressure. This lowers the rate at which a pest develops resistance to a control, because if almost all pests are killed then those that have resistance will provide the genetic basis of the future population. Retaining a significant number of unresistant specimens dilutes the prevalence of any resistant genes that appear. Similarly, the repeated use of a single class of controls will create pest populations that are more resistant to that class, whereas alternating among classes helps prevent this.

- Preventive cultural practices—Selecting varieties best for local growing conditions and maintaining healthy crops is the first line of defense. Plant quarantine and 'cultural techniques' such as crop sanitation are next, e.g., removal of diseased plants, and cleaning pruning shears to prevent spread of infections. Beneficial fungi and bacteria are added to the potting media of horticultural crops vulnerable to root diseases, greatly reducing the need for fungicides.

- Monitoring—Regular observation is critically important. Observation is broken into inspection and identification. Visual inspection, insect and spore traps, and other methods are used to monitor pest levels. Record-keeping is essential, as is a thorough knowledge target pest behavior and reproductive cycles. Since insects are cold-blooded, their physical development is dependent on area temperatures. Many insects have had their development cycles modeled in terms of degree-days. The degree days of an environment determines the optimal time for a specific insect outbreak. Plant pathogens follow similar patterns of response to weather and season.

- Mechanical controls—Should a pest reach an unacceptable level, mechanical methods are the first options. They include simple hand-picking, barriers, traps, vacuuming and tillage to disrupt breeding.

- Biological controls—Natural biological processes and materials can provide control, with acceptable environmental impact, and often at lower cost. The main approach is to promote beneficial insects that eat or parasitize target pests. Biological insecticides, derived from naturally occurring microorganisms (*e.g.—Bt*, entomopathogenic fungi and entomopathogenic nematodes), also fall in this category. Further 'biology-based' or 'ecological' techniques are under evaluation.

- Responsible use—Synthetic pesticides are used as required and often only at specific times in a pest's life cycle. Many newer pesticides are derived from plants or naturally occurring substances (*e.g.*—nicotine, pyrethrum and insect juvenile hormone analogues), but the toxophore or active component may be altered to provide increased biological activity or stability. Applications of pesticides must reach their intended targets. Matching the application technique to the crop, the pest, and the pesticide is critical. The use of low-volume spray equipment reduces overall pesticide use and labor cost.

An IPM regime can be simple or sophisticated. Historically, the main focus of IPM programmes was on agricultural insect pests. Although originally developed for agricultural pest management, IPM programmes are now developed to encompass diseases, weeds and other pests that interfere with management objectives for sites such as residential and commercial structures, lawn and turf areas, and home and community gardens.

Process

IPM is the selection and use of pest control actions that will ensure favourable economic, ecological and social consequences and is applicable to most agricultural, public health and amenity pest management situations. The IPM process starts with monitoring, which includes inspection and identification, followed by the establishment of economic injury levels. The economic injury levels set the economic threshold level. That is the point when pest damage (and the benefits of treating the pest) exceed the cost of treatment. This can also be an action threshold level for determining an unacceptable level that is not tied to economic injury. Action thresholds are more common in structural pest management and economic injury levels in classic agricultural pest management. An example of an action threshold is one fly in a hospital operating room is not acceptable, but one fly in a pet kennel would be acceptable. Once a threshold has been crossed by the pest population action steps need to be taken to reduce and control the pest. Integrated pest management employ a variety of actions including cultural controls, including physical barriers, biological controls, including adding and conserving natural predators and enemies to the pest, and finally chemical controls or pesticides. Reliance on knowledge, experience, observation and integration of multiple techniques makes IPM appropriate for organic farming (excluding synthetic pesticides). These may or may not include materials listed on the Organic Materials Review Institute (OMRI) Although the pesticides and particularly insecticides used in organic farming and organic gardening are generally safer than synthetic pesticides, they are not always more safe or environmentally friendly than synthetic pesticides and can cause harm. For conventional farms IPM can reduce human and environmental exposure to hazardous chemicals, and potentially lower overall costs.

Risk assessment usually includes four issues: 1) characterization of biological control agents, 2) health risks, 3) environmental risks and 4) efficacy.

Mistaken identification of a pest may result in ineffective actions. E.g., plant damage due to over-watering could be mistaken for fungal infection, since many fungal and viral infections arise under moist conditions.

Monitoring begins immediately, before the pest's activity becomes significant. Monitoring of agricultural pests includes tracking soil/planting media fertility and water quality. Overall plant health and resistance to pests is greatly influenced by pH, alkalinity, of dissolved mineral and oxygen reduction potential. Many diseases are waterborne, spread directly by irrigation water and indirectly by splashing.

Once the pest is known, knowledge of its lifecycle provides the optimal intervention points. For example, weeds reproducing from last year's seed can be prevented with mulches and pre-emergent herbicide.

Pest-tolerant crops such as soybeans may not warrant interventions unless the pests are numerous or rapidly increasing. Intervention is warranted if the expected cost of damage by the pest is more than the cost of control. Health hazards may require intervention that is not warranted by economic considerations.

Specific sites may also have varying requirements. E.g., white clover may be acceptable on the sides of a tee box on a golf course, but unacceptable in the fairway where it could confuse the field of play.

Possible interventions include mechanical/physical, cultural, biological and chemical. Mechanical/physical controls include picking pests off plants, or using netting or other material to exclude pests such as birds from grapes or rodents from structures. Cultural controls include keeping an area free of conducive conditions by removing waste or diseased plants, flooding, sanding, and the use of disease-resistant crop varieties. Biological controls are numerous. They include: conservation of natural predators or augmentation of natural predators, sterile insect technique (SIT).

Augmentation, inoculative release and inundative release are different methods of biological control that affect the target pest in different ways. Augmentative control includes the periodic introduction of predators. With inundative release, predators are collected, mass-reared and periodically released in large numbers into the pest area. This is used for an immediate reduction in host populations, generally for annual crops, but is not suitable for long run use. With inoculative release a limited number of beneficial organisms are introduced at the start of the growing season. This strategy offers long term control as the organism's progeny affect pest populations throughout the season and is common in orchards. With seasonal inoculative release the beneficials are collected, mass-reared and released seasonally to maintain the beneficial population. This is commonly used in greenhouses. In America and other western countries, inundative releases are predominant, while Asia and the eastern Europe more commonly use inoculation and occasional introductions.

The sterile insect technique (SIT) is an area-wide IPM program that introduces sterile male pests into the pest population to trick females into (unsuccessful) breeding encounters, providing a form of birth control and reducing reproduction rates. The biological controls mentioned above only appropriate in extreme cases, because in the introduction of new species, or supplementation of naturally occurring species can have detrimental ecosystem effects. Biological controls can be used to stop invasive species or pests, but they can become an introduction path for new pests.

Chemical controls include horticultural oils or the application of insecticides and herbicides. A green pest management IPM program uses pesticides derived from plants, such as botanicals, or other naturally occurring materials.

Pesticides can be classified by their modes of action. Rotating among materials with different modes of action minimizes pest resistance.

Evaluation is the process of assessing whether the intervention was effective, whether it produced unacceptable side effects, whether to continue, revise or abandon the program.

Southeast Asia

The Green Revolution of the 1960s and '70s introduced sturdier plants that could support the heavier grain loads resulting from intensive fertilizer use. Pesticide imports by 11 Southeast Asian countries grew nearly sevenfold in value between 1990 and 2010, according to FAO statistics, with disastrous results. Rice farmers become accustomed to spraying soon after planting, triggered by signs of the leaf folder moth, which appears early in the growing season. It causes only superficial damage and doesn't reduce yields. In 1986, Indonesia banned 57 pesticides and completely stopped subsidizing their use. Progress was reversed in the 2000s, when growing production capacity, particularly in China, reduced prices. Rice production in Asia more than doubled. But it left farmers believing more is better—whether it's seed, fertilizer, or pesticides.

The brown planthopper, *Nilaparvata lugens*, the farmers' main target, has become increasingly resistant. Since 2008, outbreaks have devastated rice harvests throughout Asia, but not in the Mekong Delta. Reduced spraying allowed natural predators to neutralize planthoppers in Vietnam. In 2010 and 2011, massive planthopper outbreaks hit 400,000 hectares of Thai rice fields, causing losses of about $64 million. The Thai government is now pushing the "no spray in the first 40 days" approach.

By contrast early spraying kills frogs, spiders, wasps and dragonflies that prey on the later-arriving and dangerous planthopper and produced resistant strains. Planthoppers now require pesticide doses 500 times greater than originally. Overuse indiscriminately kills beneficial insects and decimates bird and amphibian populations. Pesticides are suspected of harming human health and became a common means for rural Asians to commit suicide.

In 2001, scientists challenged 950 Vietnamese farmers to try IPM. In one plot, each farmer grew rice using their usual amounts of seed and fertilizer, applying pesticide as they chose. In a nearby plot, less seed and fertilizer were used and no pesticides were applied for 40 days after planting. Yields from the experimental plots was as good or better and costs were lower, generating 8% to 10% more net income. The experiment led to the "three reductions, three gains" campaign, claiming that cutting the use of seed, fertilizer and pesticide would boost yield, quality and income. Posters, leaflets, TV commercials and a 2004 radio soap opera that featured a rice farmer who gradually accepted the changes. It didn't hurt that a 2006 planthopper outbreak hit farmers using insecticides harder than those who didn't. Mekong Delta farmers cut insecticide spraying from five times per crop cycle to zero to one.

The Plant Protection Center and the International Rice Research Institute (IRRI) have been encouraging farmers to grow flowers, okra and beans on rice paddy banks, instead of stripping vegetation, as was typical. The plants attract bees and a tiny wasp that eats planthopper eggs, while the vegetables diversify farm incomes.

Agriculture companies offer bundles of pesticides with seeds and fertilizer, with incentives for volume purchases. A proposed law in Vietnam requires licensing pesticide dealers and government approval of advertisements to prevent exaggerated claims. Insecticides that target other pests, such as *Scirpophaga incertulas* (stem borer), the larvae of moth species that feed on rice plants allegedly yield gains of 21% with proper use.

The Six IPM Program Essentials

Monitoring - This includes regular site inspections and trapping to determine the types and infestation levels of pests at each site.

Record-Keeping - A record-keeping system is essential to establish trends and patterns in pest outbreaks. Information recorded at every inspection or treatment should include pest identification, population size, distribution, recommendations for future prevention, and complete information on the treatment action.

Action Levels - Pests are virtually never eradicated. An action level is the population size which requires remedial action for human health, economic, or aesthetic reasons.

Prevention - Preventive measures must be incorporated into the existing structures and designs for new structures. Prevention is and should be the primary means of pest control in an IPM program.

Tactics Criteria - Under IPM, chemicals should be used only as a last resort only, but when used, the least-toxic materials should be chosen, and applied to minimize exposure to humans and all non-target organisms.

Evaluation - A regular evaluation program is essential to determine the success of the pest management strategies.

Push–pull Agricultural Pest Management

The 'push-pull' strategy, a novel tool for integrated pest management programs, uses a combination of behavior-modifying stimuli to manipulate the distribution and abundance of insect pests and/or natural enemies. In this strategy, the pests are repelled or deterred away from the main crop (push) by using stimuli that mask host apparency or are repellent or deterrent. The pests are simultaneously attracted (pull), using highly apparent and attractive stimuli, to other areas such as traps or trap crops where they are concentrated, facilitating their control.

The term 'push-pull' was first conceived as a strategy for insect pest management by Pyke, Rice, Sabine and Zaluki in Australia in1987. They investigated the use of repellent and attractive stimuli, deployed in tandem, to manipulate the distribution of *Helicoverpa* spp. in cotton to reduce reliance on insecticides, to which the moths were becoming resistant. The concept was later formalized and refined by Miller and Cowles in the US in 1990, who termed the strategy 'stimulo-deterrent diversion' while developing alternatives to insecticides for control of the onion fly, *Delia antiqua*.

The development of a reliable, robust, and sustainable push-pull strategy requires a clear scientific understanding of the pest's biology and the behavioral/chemical ecology of the interactions with its hosts, conspecifics, and natural enemies. The specific combination of components differs in each strategy according to the pest to be controlled (its specificity, sensory abilities, and mobility) and the resource targeted for protection.

Among several push-pull strategies under development or used in practice for insect pest control, the most successful example of the push-pull strategy currently being used by farmers was developed in Africa for controlling stemborers on cereal crops. This strategy was developed using technologies appropriate to resource poor farmers and has shown a high adoption rate and spontaneous technology transfer by farmers, resulting in significant impact on food security by increased farm production in the region.

References

- van Emden, H.F.; Service, M.W. (2004). Pest and Vector Control. Cambridge University Press. p. 147. ISBN 978-0-521-01083-2
- "Augmentation: The Periodic Release of Natural Enemies". University of Wisconsin. Archived from the original on 17 March 2016. Retrieved 27 August 2017
- Shelton, A. M.; Badenes-Perez, F. R. (2005-12-06). "Concepts and applications of trap cropping in pest management". Annual Review of Entomology. 51 (1): 285–308. doi:10.1146/annurev.ento.51.110104.150959
- Wright, R. j (1984). "Evaluation of crop rotation for control of Colorado potato beetle (Coleoptera: Chrysomelidae) in commercial potato fields on Long Island". Journal of Economic Entomology. 77: 1254–1259. doi:10.1093/jee/77.5.1254
- Peter, K. V. (2009). Basics Of Horticulture. New India Publishing. p. 288. ISBN 978-81-89422-55-4. Archived from the original on 2017-04-07
- Kuris, Armand M. (March 2003). "Did biological control cause extinction of the coconut moth, Levuana iridescens, in Fiji?". Biological Invasions. 5 (1): 133–141. doi:10.1023/A:1024015327707
- "Featured Creatures. European corn borer". University of Florida IFAS. Archived from the original on 30 May 2016. Retrieved 5 June 2016
- Naranjo, Steven E. (8 June 2011). "Impacts of Transgenic Cotton on Integrated Pest Management". Journal of Agricultural and Food Chemistry. 59 (11): 5842–5851. doi:10.1021/jf102939c
- Charles Perrings; Mark Herbert Williamson; Silvana Dalmazzone (1 January 2000). The Economics of Biological Invasions. Edward Elgar Publishing. ISBN 978-1-84064-378-7
- Seaman, Abby. "Integrated Pest Management". University of Connecticut. Archived from the original on 20 February 2012. Retrieved 13 March 2012
- Clercq, P.; Mason, P. G.; Babendreier, D. (2011). "Benefits and risks of exotic biological control agents". BioControl. 56 (4): 681–698. doi:10.1007/s10526-011-9372-8
- Van Lenteren, J. C. (2009). "Implementation of biological control". American Journal of Alternative Agriculture. 3 (2–3): 102. doi:10.1017/S0889189300002265

Chapter 2

Pests: Types, Impacts and Control

A pest is an organism that is detrimental to the well being of humans, agriculture, livestock and forestry. Pests can be of two kinds, vertebrate and invertebrate. This chapter discusses in detail the different kinds of invertebrate pests, such as cereal leaf beetle, ground beetle, boll weevil, Colorado potato beetle, Japanese beetle, etc. as well as vertebrate pests like common myna, pigeon pests, rodent, mole and rabbits. It also elucidates the damage caused by such pests and the varied mechanisms to control their impect.

Pests

Pest, any organism judged as a threat to human beings or to their interests. When early man hunted animals and foraged for food, he shared the natural resources with other organisms in the community. As human culture developed and population rose, people made ever-increasing demands on these resources. One result of changing the environment has been a great increase in the number of species that are now recognized as competitors of humans. These competitors are usually referred to as pests. The definition of pest is, of course, subjective. An ecologist would not necessarily consider several leaf-eating caterpillars on a plant as pests, whereas a gardener who cultivated the plant might very well do so. And only one bat, rat, or mouse is enough to qualify as a household pest.

Natural communities have always contained organisms that were economically significant; locusts, for example, have plagued humans throughout history, and grain from Stone Age locations has been found infected with bunt and ergot diseases. Most species that became pests, however, did so because of environmental modification, occasionally from natural causes but usually from human activities.

In order to appreciate some of the methods devised to combat pests, one should consider how advancing technology has increased the number of harmful insect species. The change from natural vegetation to large areas of single-crop (monoculture) agriculture has three consequences. First, given a more uniform food source, some plant-eating species increase to large populations. Second, the uniform plant cover is easily invaded by attacking pests. Third, the introduction of new crops over large areas results in the transfer of previously harmless insects from scattered native plants to the new and abundant sources of food. Cultural practices such as fertilization, irrigation, and the use of modern harvesting equipment (which often leaves large amounts of plant litter in the field) enhance still further the ability of pest species to increase rapidly. In addition, the elimination of species that compete with or prey upon pests—an unintended effect in some pest control programs—has also exacerbated certain pest problems. Also, the ease with which people and goods can be transported around the world has resulted in the introduction of exotic pests in many places.

Pests are found throughout the animal kingdom. Microorganisms such as fungi, bacteria, and viruses are here considered with the pests even though they are usually thought of as agents of disease. Most animal pests are invertebrates, among them protozoa, flatworms, nematodes, snails, slugs, insects, and mites. Among the vertebrates, rabbits, elk, deer, and many kinds of rodents are sometimes injurious to crops.

Insects are also serious pests because some of them play an essential role in the transmission of disease. Each year millions of lives, particularly in the tropics, are threatened by insect-borne diseases. Malaria and yellow fever are transmitted by mosquitoes, plague by fleas, typhus by the human louse, sleeping sickness by tsetse flies, Chagas' disease by blood-sucking bugs, and leishmaniasis by sandflies. Other diseases may be spread by insects accidentally as a result of their habits.

Pest control began to receive attention in the 18th century, and it has become increasingly important since then. As control programs were developed, they tended to fall primarily into two somewhat mutually exclusive categories: chemical and biological. Physical or mechanical methods, including sticky barriers, heat killing (for storage pests), and flooding (for ground pests), were also developed. Because of their limited utility and short-term effectiveness, they have been superseded largely by chemical and biological methods.

The chemical, or pesticide, approach probably began with the use of poisonous plant compounds—ground tobacco was used in France to kill aphids about 1763. Other natural products such as nicotine, rotenone, petroleum, kerosene, creosote, and turpentine were used in the 19th century. Inorganic compounds such as Paris green, lime sulfur, Bordeaux mixture, hydrogen cyanide, and lead arsenate were also introduced in the 1800s.

With the appearance of the synthetic organic compounds during World War II, a dramatic change occurred in pest control. Some organic compounds, such as the dinitrophenols, had been used earlier, but the discovery of the insecticidal properties of DDT (dichloro-diphenyl-trichloroethane) and of BHC (benzene hexachloride) made the concept of pest-free crops possible. Research in the 1930s on plant hormones and related compounds led to the development of the selective herbicide 2,4-D (2,4-dichlorophenoxyacetic acid), and this became commercially available about the same time as DDT. Following the appearance of these new synthetic organic compounds, a whole new series of pesticides—insecticides, fungicides, herbicides, and plant growth regulators—was introduced.

Although the ancient Chinese used predacious ants to control foliage-feeding insects, the control of pests by biological means appears to have been initiated in the Western world with the importation of the Indian Mynah bird into Mauritius in 1762 to control the red locust. Control of bedbugs by predatory agents was recommended in 1776. A number of other projects in biological control were carried out in the 1800s, but the modern era in this phase of pest control began in 1888 with the importation into California of the vedalia beetle (Rodolia cardinalis) to control the cottony-cushion scale (Icerya purchasi). This beetle, imported from Australia, saved the citrus-fruit industry.

The use of plants resistant to insects was suggested about 1788 as a means of controlling the Hessian fly, a pest of wheat. The classic example of this approach was the control of phylloxera, aphid-like insects that attacked the root stocks of the European wine grape and almost ruined the European wine industry. The solution lay in grafting the European wine grape onto resistant American stocks. The effort, about 1900, to control the lantana shrub in Hawaii by introducing an insect appears to have been the first attempt to control weeds biologically. The use of microorganisms to destroy insect pests began in the late 1800s and early 1900s.

With the appearance of the synthetic organic insecticides, however, emphasis was placed on chemical control, and biological means fell into neglect. The extensive use of powerful chemical agents soon resulted in a number of serious ecological problems. Consequently, current pest control practice minimizes the use of pesticides and combines them with biological methods in an approach called integrated control. The integrated control of an insect that attacks crops may, for example, involve the breeding of pest-resistant crop varieties; the development of crop culture methods that inhibit pest proliferation; the release of organisms that are predators or parasites of the pest species; the placement of traps baited with the pest's own sex attractants (pheromones); the disruption of the pest's reproduction by the release of sterilized pests; and, often as a last resort, the application of chemical insecticides. Other important tools of modern pest control or importation and quarantine regulations, which are designed to prevent the introduction of exotic pests.

Types of Pests

Insects

These are important and major pests. Insects have three pairs of legs, two pairs of wings, segmented body and characteristic compound eyes and antennae. Insects are tricky and cause damage in different ways viz. sucking sap from plants, biting plant parts, boring in to fruits, twigs and leaves, attacking roots, barks and blossoms etc. The damaging stages of different insect pests are larvae, adults and nymphs.

Mites

These are creatures like insect but have soft body and four pairs of legs. These tiny creatures have red or pale yellow colour. They suck the sap from the plant and attack the crops in huge number.

Rodents

This group of pest eat away large amount of human food and also damage the crops on large scale. They are also responsible for heavy loss to stored grains on farms, in warehouses and houses.

Animals

Animals like Wild Boar, Deer, Elephants, Wild Buffalo, Jackals, Monkeys, Squirrels cause direct damage to crop plants. They eat away the plants and by and large they waste huge amount of crops.

Birds

Birds attack the crop plants and eat grains. Crow, Parrots and Sparrows are major among birds that attack the crops.

Methods of Pest Control

Pest causes lot of reduction in the yield and quality of crops. Broadly the measures of pest control are of two types, preventive and protective. Preventive measures are used before the attack of the pest and protective measures are used to control the pest after their attack. The methods of controlling pest are as follows.

Mechanical Methods

1) Picking of pests, larvae by hand and destroying them. 2) Remove the part or whole plant that is infested. 3) Use of traps. 4) Catch them with the help of net.

Physical Methods

1) By heat: High temperature kills the pest. 2) Low temperature. 3) X-rays and gamma ray.

Cultural Methods

1) Crop rotation. 2) Deep ploughing. 3) Clean cultivation. 4) Proper use of fertilizers and water. 5) Growing pest resistant Varieties. 6) Timely or late sowing. 7) Proper harvesting.

Chemical Methods

This method is used on large scale for direct control of pests. In these methods, chemicals that kill the pest are used. These chemicals are called as pesticides/ Insecticides. The chemical method is effective and faster as compared to other methods but it is hazardous to environment. Improper use of these chemical leads to development of resistance among the pests causing heavy out break which leads to increase in cost of cultivation and heavy loss. Dusts, Wettable powders, Seed dressers, Emulsions and granules are the insecticides formulations.

Biological Methods

Use of biological agents to control pests is called Bio-control. Ninety eight percent pests are controlled by nature. There are certain predators, parasites, birds, animals, micro-organisms that can be used for control of pests. Certain biological factors used to control the pests are: Tricogama, Crysopa, HNPV, LLNPV, Chilonus, Copidosoma, Bacillus thuringenesis.

Plant Quarantine Measures

Control of distribution and spread of pest by laws.

Integrated Pest Management

Owing to uncontrolled use of pesticides, the control of pest has not remained as effective as earlier. This raises resistance of pests against chemicals; increases cost of cultivation, increase residues in the produce, etc. The environmental imbalance is another effect of such methods so it is the necessity of the time to use IPM. IPM- Use of cultural, physical, mechanical, biological, environmental and legal methods collectively to control pest is called as IPM. It also uses biological extracts, traps, repellents etc. for pest control.

Invertebrate Pests

Pests such as grain mites and storage insects such as saw-toothed beetles and the grain weevil, which cause considerable damage to crops in tropical or warm temperate areas.

Insects

- Ants, cockroaches, flies and wasps are household pests, typically as they consume human food
- Aphids, caterpillars, grasshoppers and crickets cause damage to crop plants

- Lice, fleas and bed bugs can all cause skin irritation
- Mosquitoes, tsetse flies and kissing bugs cause irritation and carry disease
- Termites, silverfish, woodworm and wood ants cause structural damage
- Bookworms, carpet beetles and clothes moths cause non-structural damage
- Gypsy moths attack hardwood trees

Nematodes

- Root-knot nematode
- Soybean cyst nematode
- Potato cyst nematode

Parasites

- Chiggers cause skin irritation
- Sarcoptes scabiei causes scabies
- Ticks and mites cause irritation and can spread disease

Crustaceans

Among the crustaceans, some species of crabs from Geccarinidae and Potamonidae damage rice seedlings. Often they make holes on bands and disturb in supplying water of the crop fields.

Gastropods

Some slugs are pests in both agriculture and gardens. Their significance is increasing drastically. *Deroceras reticulatum* is a worldwide distributed slug pest. Local importance slug pests include: *Deroceras spp., Milax spp., Tandonia sp., Limax spp., Arion spp.* and some species of Veronicellidae: *Veronicella sloanei*

Land Snail Pests

Helix aspersa damages citrus fruits in California, Cernuella virgata, Theba pisana and Cochlicella spp. decrease quality of grains when harvested with the product in South Australia.

Achatina fulica damages vegetables and ornamental plants in the Pacific region. Succinea costaricana damages ornamental plants in Costa Rica. Ovachlamys fulgens damages ornamental plants and orchids in Costa Rica.

Other species considered to be pests include: Amphibulima patula dominicensis, Zachrysia provisoria and Bradybaena similaris.

Freshwater Snail Pests

Pomacea canaliculata damages rice in Southeast Asia. Bulinus sp., Biomphalaria spp. and Oncomelania are intermediate hosts of schistosomes causing schistosomiasis.

Various species in Lymnaeidae are intermediate hosts of fasciolosis.

Other Invertrebate Pests

Armyworms can be problematic to grasses and crops at this time of year. They sometimes feed on the remaining green material just below the maturing head of cereals, causing heads to fall. Ripening barley crops are generally worst affected. The first visible sign of armyworm caterpillars is often their green to straw-coloured droppings, about the size of a match head, found on the ground between the cereal rows. Damage to weeds, especially their preference for ryegrass, is also a sign of their presence. Armyworms may be confused with loppers, budworm larvae, cabbage moth larvae and cutworms. Armyworm caterpillars are fat and smooth, and may be distinguished by the three parallel white stripes on the collar just behind their head.

Bryobia mites are most active in warm conditions in autumn, spring and summer. At this time of year, bryobia mite populations can build up very quickly. They attack clovers, lucerne, lupins and canola. The feeding damage is characterised by a long trail of whitish-grey spots on the upper side of cotyledons and leaves. In the field, they are often mistaken for redlegged earth mites. Bryobia mites, often called clover mites, can be distinguished by their long forelegs which are 1.5 times the body length.

Cereal Leaf Beetle

Oulema Melanopus (L.)

Cereal leaf beetle is an occasionally severe pest of wheat and oats in Pennsylvania, but it also feeds on barley, rye, and other grasses. This insect species is native to Europe and Asia and was first detected in the United States in 1962, but quickly spread to most of the wheat-growing areas of the eastern United States. This pest has been known to catch small-grain growers off guard because it tends not to be a perennial problem; however, it can be extremely damaging if populations build and infestations are not detected early. It is more of a problem in thin or poorly established stands, so it can often be controlled effectively just by using good crop management practices. In many parts of its range, it is controlled by introduced species of parasitic wasps, but these wasps may not be well established in Pennsylvania and some states to south.

Description

Adult cereal leaf beetle.

Adult cereal leaf beetles (figure above) are about 5-mm long (3/16 inch) and have metallic bluish-black heads and elytra (i.e., wing covers). Their thorax and legs are orange to reddish brown. Eggs are about 1-mm (< 1/16 inch) long and yellow and typically laid singly or in pairs on the upper surface of grass leaves. Eggs darken and turn black as hatching approaches. Larvae are yellowish-orange, but this color is usually obscured by a layer of feces and mucus, giving them a shiny appearance. This layer protects them from natural enemies and drying out and has been known to ruin a favorite pair of trousers as fields with heavy infestations are walked.

Life History

Cereal leaf beetle has one generation per year. Adults pass the winter in leaf litter in wood lots, hedgerows, and similar areas. In March and mid-April as temperatures increase, adults fly to grain fields to lay eggs, which hatch and young larvae begin to feed. Young larvae do not inflict much damage, but older larvae have voracious appetites and can cause extensive damage (figure below). The larval stage passes in about two weeks depending on spring temperatures. Mature larvae burrow into the ground to pupate. Adults emerge about two weeks later, and will feed in small-grain and corn fields for a short time before remaining inactive for most of the summer. In fall, adults fly to wooded areas seeking shelter for winter.

Damage

Older larvae do the great majority of damage, stripping off green tissue between leaf veins and skeletonizing leaves. Large populations can cause fields to turn white; indeed, this "frosted" appearance is often what tips growers off to the presence of damaging populations in their fields (figure below). Adult feeding in small grains and corn has not been shown to be economically significant and targeting adults for control has not been shown to be effective.

Cereal leaf beetle larvae feeding on wheat

A "frosted" wheat field resulting from heavy infestation of cereal leaf beetle.

Management

This pest is most easily managed earlier in spring when eggs or young larvae are present; therefore, in-depth, localized scouting is necessary to detect these life stages. Keep in mind that populations can be quite spotty so checking individual fields is often necessary. Once populations reach one or more larvae per stem, treatment is probably warranted, especially if larvae are feeding on the flag leaf prior to head emergence. Damage later in head filling does not appear to be that significant and if adults are seen late in spring, it is likely too late to manage this pest.

Ground Beetle

Ground beetle, (family Carabidae), any member of more than 40,000 insect species in one of the largest families in the insect order Coleoptera. They can be found in almost any terrestrial habitat on Earth. Ground beetles are recognized by their long legs and shiny black or brown elytra (wing covers), which are decorated with ridges and may be fused together along the midline. In many species the hind wings are reduced or absent. Ground beetles prefer moist cool areas and usually run rather than fly when disturbed. They emerge from under rocks, crevices, or litter at night in search of insects, worms, or snails. The long, slender larvae are mostly carnivorous, although those of a few species feed on seeds. They have sharp projecting mouthparts and a pair of bristly tail appendages. Many ground beetles secrete a foul-smelling liquid that discourages potential predators such as birds.

The searcher, or caterpillar hunter (*Calosoma scrutator*), is a common, brightly coloured North American ground beetle about 35 mm (1.5 inches) long. Its green or violet wings are edged in red, and its body has violet-blue, gold, and green markings. This and related species of ground beetles are known to climb trees in search of caterpillars. They secrete an acidic fluid that can blister human skin. The iridescent green-coloured *C. sycophanta* was introduced into North America from Europe to help control the gypsy moth and brown-tail moth caterpillars.

The snail hunters (e.g., *Scaphinotus*) are a specialized group of ground beetles. Elongated, hook-shaped mouthparts allow them to extract the snail from its shell. The bombardier beetle (*Brachinus*in North America and *Pherosophus* in Africa, Asia, and the East Indies) has little sacs at the tip of its abdomen that spray a noxious fluid used to deter enemies. The beetle fires this fluid at boiling-hot temperatures, and the hot fluid vaporizes upon contact with air. The fluid itself consists of toxins called quinones that react explosively in an internal chamber with hydrogen peroxide, which is also produced by the beetle and stored in a separate body compartment. The expulsion of fluid from the abdomen is accompanied by a distinctive popping sound that may further startle the beetle's predators. *Brachinus* species have dark blue, black, or blue-green wings and reddish yellow bodies and legs.

The beneficial *Lebia grandis,* which resembles the bombardier beetle, preys upon the Colorado potato beetle. The Malayan leaf beetle, or fiddle beetle (*Mormolyce*), measuring approximately 100 mm (4 inches) long, resembles a violin with its slender head and thorax and wide elytra. This flat beetle uses its long head to probe into small openings in search of prey. It hides in crevices, under tree bark, and in porous fungi. The majority of ground beetles feed on insects harmful to humans and are therefore considered to be beneficial.

Description and Ecology

Although their body shapes and coloring vary somewhat, most are shiny black or metallic and have ridged wing covers (elytra). The elytra are fused in some species, particularly large Carabinae, rendering the beetles unable to fly. The genus *Mormolyce* is known as violin beetles due to their peculiarly shaped elytra. All carabids except the quite primitive flanged bombardier beetles (Paussinae) have a groove on their foreleg tibiae bearing a comb of hairs used for cleaning their antennae.

A *Brachinus* species typical bombardier beetle (Brachininae: Brachinini) from North Carolina

Defensive Secretions

Typical for the ancient beetle suborder Adephaga to which they belong, they have paired pygidial glands in the lower back of the abdomen. These are well developed in ground beetles, and produce noxious or even caustic secretions used to deter would-be predators. In some, commonly known as bombardier beetles, these secretions are mixed with volatile compounds and ejected by a small combustion, producing a loud popping sound and a cloud of hot and acrid gas which can injure small mammals, such as shrews, and is liable to kill invertebrate predators outright.

To humans, getting "bombed" by a bombardier beetle is a decidedly unpleasant experience. This ability has evolved independently twice as it seems – in the flanged bombardier beetles (Paussinae) which are among the most ancient ground beetles, and in the typical bombardier beetles (Brachininae) which are part of a more "modern" lineage. The Anthiini, though, can mechanically squirt their defensive secretions for considerable distances and are able to aim with a startling degree of accuracy; in Afrikaans, they are known as *oogpisters* ("eye-pissers"). In one of the very few known cases of a vertebrate mimicking an arthropod, juvenile *Heliobolus lugubris* lizardsare colored similar to the aposematic *oogpister* beetles, and move in a way that makes them look surprisingly similar to the insects at a casual glance.

A folk story claims that Charles Darwin once found himself on the receiving end of a bombardier beetle's attack, based on a passage in his autobiography. Darwin stated in a letter to Leonard Jenyns that a beetle had attacked him on that occasion, but he did not know what kind:

A *Cychrus rostratus* once squirted into my eye & gave me extreme pain; & I must tell you what happened to me on the banks of the Cam in my early entomological days; under a piece of bark I found two carabi (I forget which) & caught one in each hand, when lo & behold I saw a sacred *Panagæus crux major*; I could not bear to give up either of my

Carabi, & to lose *Panagæus* was out of the question, so that in despair I gently seized one of the carabi between my teeth, when to my unspeakable disgust & pain the little inconsiderate beast squirted his acid down my throat & I lost both Carabi & *Panagæus*!

A *Lophyra* sp. tiger beetle from Tanzania

Ecology

Common habitats are under the bark of trees, under logs, or among rocks or sand by the edge of ponds and rivers. Most species are carnivorous and actively hunt for any invertebrate prey they can overpower. Some run swiftly to catch their prey; tiger beetles (Cicindelinae) can sustain speeds of 9 km/h – in relation to their body length they are among the fastest land animals on Earth. Unlike most Carabidae which are nocturnal, the tiger beetles are active diurnal hunters and often brightly coloured; they have large eyes and hunt by sight. Ground beetles of the species *Promecognathus laevissimus* are specialised predators of the cyanide millipede *Harpaphe haydeniana*, countering the hydrogen cyanide which makes these millipedes poisonous to most carnivores.

Relationship with Humans

As predators of invertebrates, including many pests, most ground beetles are considered beneficial organisms. The caterpillar hunters (*Calosoma*) are famous for their habit of devouring prey in quantity, eagerly feeding on tussock moth (Lymantriidae) caterpillars, processionary caterpillars (Thaumetopoeidae) and woolly worms(Arctiidae), which due to their urticating hairs are avoided by most insectivores. Large numbers of the forest caterpillar hunter (*C. sycophanta*), native to Europe, were shipped to New England for biological control of the gypsy moth (*Lymantria dispar*) as early as 1905.

A few species are nuisance pests. *Zabrus* is one of the few herbivorous ground beetle genera, and on rare occasions *Zabrus tenebrioides*, for example, occurs abundantly enough to cause some damage to grain crops. Large species, usually Carabinae, can become a nuisance if present in numbers, particularly during outdoor activities such as camping; they void their defensive secretions when threatened, and in hiding among provisions their presence may spoil food. Since ground beetles are generally reluctant or even unable to fly, mechanically blocking their potential routes of entry is usually easy. The use of insecticides specifically for Carabid intrusion may lead to unfortunate side effects, such as the release of their repugnatorial secretions, so it generally is not a good idea unless the same applications are intended to exclude ants, parasites or other crawling pests.

Crucifix ground beetle (*Panagaeus cruxmajor*)

Especially in the 19th century and to a lesser extent today, their large size and conspicuous coloration, as well as the odd morphology of some (e.g. the Lebiini), made many ground beetles a popular object of collection and study for professional and amateur coleopterologists. High prices were paid for rare and exotic specimens, and in the early to mid-19th century, a veritable "beetle craze" occurred in England. As mentioned above, Charles Darwin was an ardent collector of beetles when he was about 20 years old, to the extent that he would rather scour the countryside for rare specimens with William Darwin Fox, John Stevens Henslow, and Henry Thompson than to study theology as his father wanted him to do. In his autobiography, he fondly recalled his experiences with *Licinus* and *Panagaeus*, and wrote:

No poet ever felt more delight at seeing his first poem published than I did at seeing in Stephen's *Illustrations of British Insects* the magic words, "captured by C. Darwin, Esq."

Evolution and Systematics

The Adephaga are documented since the end of the Permian, about 250 million years ago. Ground beetles evolved in the latter Triassic, having separated from their closest relatives by 200 million years ago. The family diversified throughout the Jurassic, and the more advanced lineages, such as the Harpalinae, underwent a vigorous radiation starting in the Cretaceous. The closest living relatives of the ground beetles are the false ground beetles (Trachypachidae) and the wrinkled bark beetles (Rhysodidae). They are sometimes even included in the Carabidae as subfamilies or as tribes *incertae sedis*, but more preferably they are united with the ground beetles in the superfamily Caraboidea.

Much research has been done on elucidating the phylogeny of the ground beetles and adjusting systematics and taxonomy accordingly. While no completely firm consensus exists, a few points are generally accepted: As it seems, the ground beetles consist of a number of more basal lineages and the extremely diverse Harpalinae which contain over half the described species and into which several formerly independent families had to be subsumed.

Basal Ground Beetles

Carabus lateralis (Carabinae: Carabini)

Notiophilus substriatus (Nebriinae: Notiophilini)

Clivina fossor (Scaritinae: Clivinini)

Amblytelus sp. (Amblytelinae: Amblytelini)

Carabinae Latreille, 1802 – including Agoninae and Callistinae

- Altagonum
- Aplothorax (monotypic genus)
- Calosoma - including Callisthenes
- Carabus
- Cychrus
- Homothes
- Fortagonum
- Laemostenus
- Notagonum
- Coptolabrus

Cicindelinae – tiger beetles (roughly 2,600 species; sometimes included in the Carabidae)

Cicindinae

Elaphrinae Latreille, 1802

- Blethisa
- Diacheila Motschulsky, 1844
- Elaphrus

Hiletinae

Loricerinae Bonelli, 1810

- Loricera

Migadopinae

Nebriinae (includes Notiophilinae, often included in Carabinae)

- Leistus
- Nebria
- Notiophilus
- Pelophila Dejean, 1821
- Nippononebria

Nototylinae

Omophroninae Bonelli, 1810 – round sand beetles

- Omophron

Paussinae – ant nest beetles, flanged bombardier beetles

Promecognathinae

Scaritinae Bonelli, 1810 – pedunculate ground beetles

- Clivina
- Dyschirius Bonelli, 1810
- Scarites

Siagoninae Bonelli, 1810

Carabidae Conjunctae

Dixus sphaerocephalus (Harpalinae: Harpalini)

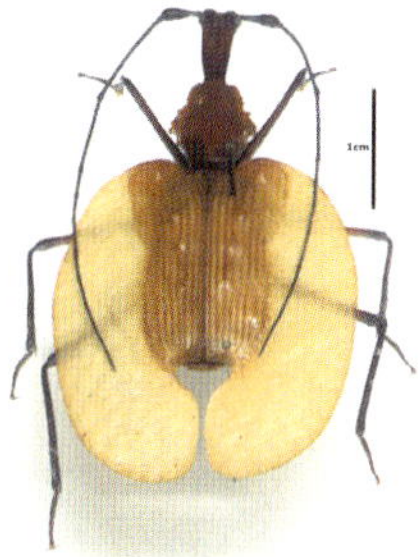

Mormolyce phyllodes (Lebiinae: Lebiini)

Amblytelinae Sloane, 1898

- *Amblytelus*

Apotominae

Brachininae – typical bombardier beetles

Broscinae

Dryptinae (sometimes in Harpalinae)

Gineminae (sometimes in Harpalinae)

Harpalinae (over 20,000 species)

Lebiinae – including Cyclosominae, Mormolycinae, Odacanthinae, Perigoninae (sometimes in Harpalinae)

Licininae – including Chlaeniinae, Oodinae (sometimes in Harpalinae)

Melaeninae

Migadopinae

Orthogoniinae (sometimes in Harpalinae)

Panagaeinae (sometimes in Harpalinae)

Platyninae (sometimes in Harpalinae)

Pseudomorphinae (sometimes in Harpalinae)

Psydrinae

Pterostichinae – including Zabrinae (sometimes in Harpalinae)

Trechinae – including Bembidiinae, Patrobinae

Tribes *Incertae Sedis*

- Amarotypini – Migadopinae or a distinct subfamily?
- Gehringiini – Psydrinae, Trechinae or a distinct subfamily?

Problems caused by Ground Beetles

As beneficial insects that play an important role in curbing populations of pests and weeds, ground beetles generally only become a problem when they occur in large numbers and move indoors. A typical ground beetle infestation comprises just a handful of the harmless insects. In some cases, however, ground beetles will enter buildings in large numbers and cause panic or alarm, even though the pests do not pose a health threat to people or damage property. Certain species are capable of releasing odorous secretions, while other types of ground beetles can be major crop pests that feed on seeds of corn.

Detection / Signs of Infestation

Ground beetles are occasional invaders, they are typically encountered from spring to fall and enter homes most frequently during the middle or end of summer. Attracted to light, ground beetles are noticeable when they congregate around well-lit structures, before entering all at once and creating an infestation. Since ground beetles do not breed indoors, they leave behind few other signs of an infestation.

Prevention Tips

Ground beetles often gain access to buildings through cracks, crevices, and other small openings, repairing and sealing these potential entry points will help keep the pests from invading. Other ways to prevent a ground beetle infestation include keeping mulch away from the building foundation and either reducing the level of outdoor lighting within the immediate vicinity of the structure or using yellow light bulbs instead of bright white ones. Furthermore, the yard should remain free of rocks, stones, wood, boards, leaves, and other debris under which ground beetles typically like to live.

Control / Removal

Smaller infestations of ground beetles can often be vacuumed up or captured in a container and removed. Sticky traps may also help control and manage the pests. Large infestations, on the other hand, may require the application of an appropriately labelled insecticide around the perimeter of the home. Insecticides must be handled carefully and applied in accordance to label directions for safety and in order to work effectively. A knowledgeable pest control professional can be consulted to manage an infestation while providing additional information on how to prevent ground beetles from invading in the future.

Boll Weevil

Boll weevil, (*Anthonomus grandis*), beetle of the insect family Curculionidae (order Coleoptera), a cotton pest in North America. Introduced to the United States from Mexico in the 1890s, the boll weevil was a severe agricultural pest for nearly 90 years, until the launch of an aggressive multiyear eradication campaign in 1978. The campaign almost, progressing slowly but effectively, completely eradicated the boll weevil from cotton-producing states, primarily through aerial release of the insecticide malathion. The boll weevil infestation is estimated to have caused crop losses of 30 to 50 percent in infested areas. The eradication program led to increased crop yields (by 10 percent or more) and a dramatic decrease in the use of insecticides (40–100 percent), leading to a reduction in production costs. In 2013 Texas was the only state to still have areas with boll weevil infestations.

Boll weevil (*Anthonomus grandis*)

The size of a mature boll weevil varies according to the amount of food it receives during its larval stage, but it averages about 6 mm ($^{1}/_{4}$ inch), including the long, curved snout, which is about one-half the body length. In the spring, adult boll weevils emerge from a partly dormant state, and their light yellow colour changes to gray or black over several weeks. Females deposit between about 100 and 300 eggs in cotton buds or fruit, though they avoid depositing their eggs in cotton bolls already visited by other females, unless most of the bolls are infested. An average of two to three weeks is required for an egg to develop into an adult, and there may be up to 10 generations each year. The larvae live entirely within the cotton boll, destroying not only the seeds but also the surrounding cotton fibres. Because the larvae and pupae remain inside the cotton bolls for their entire period of development, the application of insecticides at that time is ineffective.

The boll weevil infestation caused many farmers to realize the value of crop rotation and the need for crop diversification rather than total dependence on cotton. In addition to the use of malathion, control programs include early destruction of cotton stalks, cleanup of hibernating areas, seed treatments, early planting, and the development of early-maturing and rapid-fruiting varieties of cotton.

Boll Weevil Control Methods

To have satisfactory control is dependent on using an integrated management system that includes the use of cultural and chemical methods.

Cultural Control

It is recommended that cultural control practices include early planting, stimulating fast and efficient growth by properly prepared seed beds, adequate fertilization, weed control practices and planting early maturing adapted varieties of cotton. Due to the large infestation of boll weevil and it's difficultly to control, it is important to have good crop rotation and crop diversification. Other good cultural control programs include early detection of cotton stalks, cleanup of ears that are in hibernation of boll weevils, early planting, seed treatment and cotton varieties that are early to mature and have rapid-fruiting. The main objective of these cultural control practices is to rush the development of cotton plants so the plants are set before boll weevils become established and abundant.

Chemical Control

The timing of chemical control of boll weevil is very important. Due to the fact that the boll weevil larvae spend their entire developing period inside cotton bolls, the application of insecticides would be greatly ineffective at this time. For optimal insecticidal control would include both an in-season and diapause application. In-season application should be based on weekly boll weevils counts and damage and is done to control the weevils during the major period of boll maturity and fruit set. The first application of insecticides should be done when 10% of the squares are punctured. For the most part, insecticides are used to reduce the overwintering of boll weevils. This practice is used to delay the need for in-season application the following year. When needed, treatments should start at the beginning of diapause and continue until the fields no longer have food for boll weevil. Application of a chemical called defoliant near the end of the season speeds up harvesting and damages the crop residue as early as possible, resulting in the eradication of the food source for the remaining boll weevils.

Colorado Potato Beetle

The Colorado potato beetle, *Leptinotarsa decemlineata* (Say), is a major pest of potato, *Solanum tuberosum* L., in commercial production and home gardens in Minnesota. Larvae and adults both feed on foliage and if left untreated, complete defoliation of plants (figure below) is possible. The Colorado potato beetle was first recognized as a pest of potato in Colorado in 1859 after settlers introduced potatoes into the insect's native range of the eastern Rocky Mountains. The native host for this insect is a relative of potato, buffalo bur (*Solanum rostratum*). It took the beetle about 30 years to adapt to potato. But, once Colorado potato beetles had adapted to feeding on potato, the beetles migrated to the east using potatoes grown in farms and gardens throughout the Great Plains and Ohio River Valley. On average, the Colorado potato beetle expanded its range eastward approximately 85 miles per year, reaching the East Coast by 1874. The insect has since become established in Europe and spread eastward to the Urals and Turkey. Potato growers have struggled to control this insect since 1865 when the first broad-spectrum insecticide, Paris green (lead arsenate), was dusted onto potato leaves to protect the foliage from Colorado potato beetles.

Biology and Life Cycle

Adult Colorado potato beetles (figure above) are about half the size of your thumbnail, oval in shape, and are a yellow-orange color with 10 narrow black stripes on the forewings (elytra). The adult Colorado potato beetle overwinters in the soil either in the potato field itself or in field margins. In Minnesota, the beetle overwinters most successfully in windbreaks and other wooded areas surrounding potato fields. The following spring, adults begin to emerge (figure below) at about the same time potatoes emerge. Adults feed for a short while in the spring, then begin to mate and lay clusters of 10 – 30 yellow eggs on the underside of the leaf (figure below). Females typically lay 350 or more eggs during their lives that can last several weeks. The Colorado potato beetle has few natural enemies, and those that do feed on the eggs, larvae, pupae, or adults have little impact on the Colorado potato beetle populations.

Larval Development

Eggs hatch within 1-2 weeks, depending upon temperatures (table below). Larvae remain aggregated near the egg mass when young (figure below) but begin to move throughout the plant as that leaf is consumed (figure below). Larvae molt four times getting progressively larger each time. Over 75% of the defoliation caused by the immatures occurs during the last (4th) larval instar (figure below). Young larvae are brick red with a black head capsule while 3rd and 4th instar larvae are pink to salmon colored with a black head capsule. The sides of the larvae have two rows of dark spots. Larvae can complete development in as little as 8-10 days if the average temperatures are in the mid 80's while it will take over a month if temperatures average near 60 F. The fourth instar larvae eventually drop from the plant, and burrow into the soil to pupate.

Table: Colorado potato beetle development (days) in relation to average daily temperatures.

Average Daily Temperature (F)	Days for eggs to hatch	Days to complete larval development	Days to complete pupation	Days for Summer adults to appear
	12	28	*	*
65	8	19	18	45
70	6	14	14	34
75	5	11	11	27
80	4	9	9	22
85	3.5	8	8	19.5

* No insects pupated at these temperatures. Adapted from Vegetable Insect Management, Meister Publishing.

Damage

Both adults and larvae are voracious herbivores. Typically young larvae begin feeding on newly expanding foliage at the top of the plant. Though defoliation may seem extensive on young plants, this early feeding damage on newer leaves is not damaging to yield at harvest. As larvae mature they typically feed in groups first defoliating whole leaves then ultimately the entire plant. The third and fourth instars do the majority of feeding among all stadia. Larvae will on average consume 40 cm^2 of potato tissue while adults will consume an additional 9.65 cm^2.

Scouting

Begin checking for CPB adults in early May after potato plants have emerged. Early detection of infestation is especially critical in newly transplanted, susceptible eggplant crops. Visual search patterns for CPB should be twofold early in the growing season. First, quickly scan plants and surrounding soil for the presence of live adults. Often adult beetles will drop from small plants to the soil as a defensive tactic, typically observing the area surrounding plants will help for early detection. Second, carefully examine lower leaf surfaces of plants for clusters of bright yellow-orange, waxy eggs. Note the number of adults and egg masses in for a given number of plants. This number will help to track the overall trajectory of pest infestation. Additionally, focus early season scouting on border rows that are adjacent to either previous solanaceous crops or unmanaged non-crop areas. These have the greatest probability for early infestation by adult CPB and greater densities egg masses.

Degree-day Model

Table: Rate of beetle development using degree days

Life stage	Degree days	Accumulated degree days	Treatment
Egg	120	120	Not susceptible – do not treat
First instar	65	185	Most effective time to apply Btt
Second instar	55	240	Most effective time to apply conventional insecticides
Third instar	60	300	Most effective time to apply conventional insecticides
Fourth instar	100	400	Most effective time to apply conventional insecticides
Pupae	275	675	Not susceptible – do not treat

Most chemical pest management practices are timed to calendar dates. But insect development is directly related to temperature: cool weather slows growth, warm weather accelerates it. So unless the weather of a given year is that of the "normal" year, recommended treatment dates may not coincide with the most vulnerable life stage of an insect pest. Using a system based on the daily high and low temperatures instead of calendar dates will help you better anticipate pest outbreaks. This system converts daily average temperatures into degree days. A degree day is a unit of measure for each degree above a base temperature (52°F for CPB). To use this approach, begin keeping track of the temperature when you find the first egg mass. Keep a running total of the numbers of degree days to chart insect development. Table above identifies the number of degree days needed for each stage of Colorado potato beetle development.

After the first detection of either adults or egg masses begin strategically looking for small larvae (1st and 2nd instars) on the crop. To do this effectively, focus the majority of scouting for larvae on the tops of plants. In potato, small larvae will typically move to the newly expanding foliage at the crown of the plant. Larvae are very small at this stage and may not be directly apparent unless the emerging foliage is carefully parted. First instar larvae may be easily confused with dirt or debris until closer inspection. This task is even more challenging on some varieties of eggplant due to the darker colored foliage and stems. Careful inspection of the crop at this stage greatly enhances your chances for season long management of CPB. As the larvae grow larger general defoliation will first be apparent as small holes forming on upper leaves. Potato beetle larvae are typically very immobile and should be in close proximity to feeding damage.

Secondary pests such as the highly mobile potato flea beetle (*Epitrix cucumeris*) may also cause feeding damage to both potato and eggplant foliage. If potato beetle larvae are not present when scouting foliage there is a good chance the feeding may be caused by potato flea beetle adults. Flea beetle damage often appears in a shotgun pattern (several small holes in leaves) that differs from skeletonizing feeding patterns of CPB. Once CPB are large larvae (3rd and 4thinstars) defoliation will be much more apparent. Entire leaves will be stripped of all foliage throughout the plant. CPB populations often become asynchronous as the season progresses. It is very important to continue searching for all life stages throughout the season. Careful and deliberate scouting season-long will go a long way to gain the upper hand on CPB.

Management

Crop Tolerance

Potato plants can tolerate varying levels of defoliation before they will suffer yield losses. The level of tolerance depends on the plant's growth stage. Flowering plants can tolerate the least defoliation, only 5-10% of total leaf area. Post-flower potato is able to withstand a slightly greater amount of defoliation but since this is a critical point for tuber formation and bulking growers should limit the amount of feeding done by CPB. Growth stages ahead and behind these two points are more tolerant of defoliation. Late season feeding on potato that is beginning to senesce is the least critical period for yield loss. Eggplant is far less tolerant of defoliation, growers should be mindful of managing infestations carefully to limit complete crop loss to CPB. It should be noted that estimation of crop defoliation is an exceptionally challenging task for most growers. Defoliation estimates are typically 10-20% higher than true leaf area eaten by CPB. To ensure a good representation of potential crop damage integrates both insect counts and defoliation for an estimate of current and future damage to the crop.

Biological Control

Exceptionally high fecundity of CPB makes them very difficult to manage with natural enemy pressure alone. In the diverse polyculture of the home gardens and fresh market operations this strategy may be more effective. *Beauveria bassiana* is a natural, soil-dwelling pathogenic fungus known to infect a wide range of insects. It is available from many garden stores as a formulated compound that can easily be applied as a foliar spray. In some cases *B. bassiana* has been shown to reduce CPB populations upto 75%.

Several predaceous bugs, beetles, two species of tachinid flies, and generalist spider predators have been shown to feed on CPB. A small parasitic wasp (*Edovum puttleri*) has been documented to parasitize 71-91% of CPB eggs in eggplant, effectively killing the majority of eggs contacted. Though, some of these natural enemies show promise in small-scale studies, few have demonstrated success in large-scale field situations.

Cultural and Mechanical Control

A variety of cultural control methods are available to reduce CPB. In home garden, the most effective way to manage beetles is by hand. When scouting plants, gardeners can easily crush egg masses and groups of small larvae with their fingers. Hand picking large larvae and adult CPB into a cup of soapy water is a very easy method to reduce older life stages. Be sure to wash your hands before touching your eyes or mouth after hand picking CPB because the insects contain a chemical that is often irritable to sensitive skin. Hand picking beetles is very effective but growers must be persistent to keep populations under control. Be sure to check plants every few days until the peak colonization period has ended.

Hand picking is a challenge in many larger fresh-market gardens, so other strategies are needed. Crop rotations that avoid solanaceous crops such as eggplant or potato will substantially reduce the beetle population. This practice must be incorporated with non-crop sanitation of other weedy solanaceous hosts for CPB. Another effective management strategy is to plant extra rows of potatoes adjacent to overwintering sites to aggregate beetles in the spring and autumn. Beetles will are attracted to these "trap" crops and can be removed using propane flamers, vacuum suction, flail chopping or other physical controls. Spring trap crops must be planted a full two weeks prior to the primary crop for greatest efficacy, relying upon early emerging potato being more attractive to colonizing adults.

Physical barriers, such as floating row covers and plastic-lined trenches, can be used to prevent adult beetles from finding plants and laying eggs. Row covers should be in place before plants emerge and should remain there through the peak colonization period (mid-June). Plastic-lined trenches should be situated between fields and overwintering sites before plants emerge. Dig trenches 50 cm deep and 30 cm wide, line them with plastic, and allow plastic to become covered with a thin layer of dust from field soil. Beetles will become trapped in the trench and will be unable to climb out.

Chemical Control

Colorado potato beetle has a long history of insecticide resistance and this remains a major pest of commercial potato production. Wisconsin beetle populations are still susceptible to many insecticides, but repeated applications and overuse of similar insecticide classes will lead to insecticide resistance issues.

For most CPB chemical management tools, timing application occurs with the appearance of most first instar larvae in the field. Early instar larvae are the most susceptible life stage for chemical management, applications should be timed with the beginning of egg hatch. The first application should be followed up in 5-7 days with a second application of the same compound depending on the formulation and label restrictions.

Japanese Beetle

Japanese beetles are small bugs that carry a big threat. They do not discriminate on what types of plants they feed on. In fact, they are classified as a pest to hundreds of different species. They are one of the major insect pests in the Eastern and Midwestern United States, causing monumental damage to crops each year.

Prior to the beetle's accidental introduction into the United States in the early 1900s, the Japanese beetle was found only on the islands of Japan, isolated by water and kept in check by its natural enemies. In 1912, a law was passed that made it illegal to import plants rooted in soil. Unfortunately, the failure to implement the law immediately allowed the Japanese beetle to arrive in this country.

Most entomologists agree that the beetles entered the country as grubs in soil on Japanese iris roots. In 1916, these coppery-winged pests were first spotted in a nursery near Riverton, New Jersey. By 1920, eradication programs were dropped; the beetle proved to be too prolific a breeder. Not a choosy eater, it dines on over 200 species of plants.

Identification

How to Identify Japanese Beetles

Japanese Beetles are ½ inch in length with metallic blue-green heads, copper backs, tan wings, and small white hairs lining each side of the abdomen. Japanese beetles usually feed in small groups. They lay eggs in the soil during June, which develop into tiny white grubs with brown heads and six legs that are up to ¾ inch in length. These grubs will remain under wraps for about 10 months, overwintering and growing in the soil.

They emerge from the soil as adult beetles and begin feeding in June. They usually attack plants in groups, which is why damage is so severe. Although the lifecycle of the adult Japanese beetle is barely 40 days, it can cover a lot of ground. Even if you succeed in controlling your Japanese beetle population, your neighbor's Japanese beetles might come on over.

Japanese Beetle Damage

Japanese beetles eat a wide variety of flowers and crops (the adult beetles attack more than 300 different kinds of plants), but they are especially common on roses, as well as beans, grapes, and raspberries.

They can devour most of the foliage on favored plants like roses. Look for leaves that are "skeletonized" (only have veins remaining). This is a tell-tale sign of Japanese Beetles. Mexican Bean Beetles can also leave foliage skeletonized, though, so be sure to identify them by their appearance as well.

Grubs damage grass when overwintering in the soil, as they eat the roots of lawn grasses and garden plants.

Japanese beetles cause leaves to appear skeletonized

Control and Prevention

How to Get Rid of Japanese Beetles

Good horticultural practices, including watering and fertilizing, will reduce the damage caused by these beetles, but oftentimes you simply need to get rid of them. Here are some ideas:

- Row Covers: Protect your plants from Japanese beetles with row covers during the 6- to 8-week feeding period.

- Hand Pick: Unfortunately, the most effective way of getting rid of Japanese beetles is to hand pick them. It's time consuming, but it works, especially if you are diligent. When you pick them off, put them in a solution of 1 tablespoon of liquid dishwashing detergent and water, which will cause them to drown.
- Neem Oil: Neem oil and sprays containing potassium bicarbonate are somewhat effective, especially on roses. The adults ingest a chemical in the neem oil and pass it on in their eggs, and the resulting larvae die before they become adults. Neem can be harmful to fish and should be reapplied after rainstorms.
- Use a Dropcloth: Put down a dropcloth and, in the early morning when they're most active, shake them off and dump them into a bucket of soapy water.
- Insecticides: If you wish to spray or dust with insecticides, speak to your local cooperative extension or garden center about approved insecticides in your area.
    1. Or, try this safe homemade solution: Mix 1 teaspoon of liquid dishwashing detergent with 1 cup of vegetable oil and shake well; then add it to 1 quart of water. Add 1 cup of rubbing alcohol and shake vigorously to emulsify. Pour this mixture into a spray bottle and use it at ten-day intervals on pests. Homemade sprays can run more of a risk of damaging plant leaves, so be careful.
    2. Apply sprays in the morning, never in full sun or at temperatures above 90°F. If your plants start to wilt, rinse the leaves immediately with clean water.
- Traps: Japanese beetle traps can be helpful in controlling large numbers of beetles, but they also might attract beetles from beyond your yard. Eugenol and geraniol, aromatic chemicals extracted from plants, are attractive to adult Japanese beetles as well as to other insects. Unfortunately, the traps do not effectively suppress adults and might even result in a higher localized population. If you want to try them, be sure to place traps far away from plants so that the beetles do not land on your favored plants on their way to the traps.
- Fruit Cocktail: You can buy Japanese beetle traps of all sorts, but most are no more effective than a can of fruit cocktail. Open the can and let it sit in the sun for a week to ferment. Then place it on top of bricks or wood blocks in a light-colored pail, and fill the pail with water to just below the top of the can. Place the pail about 25 feet from the plants you want to protect. The beetles will head for the sweet bait, fall into the water, and drown. If rain dilutes the bait, start over.
- Geraniums: Japanese beetles are attracted to geraniums. They eat the blossoms, promptly get dizzy, fall down, and permit you to dispose of them conveniently with a dustpan and brush. Plant geraniums close to more valuable plants which you wish to save from the ravages of Japanese beetles.
- Japanese Beetles on Roses? Note that insecticides will not fully protect roses, which unfold too fast and are especially attractive to beetles. When beetles

are most abundant on roses, nip the buds and spray the bushes to protect the leaves. When the beetles become scarce, let the bushes bloom again. Timeliness and thoroughness of application are very important. Begin treatment as soon as beetles appear, before damage is done.

Western Corn Rootworm

Corn rootworms are important insect pest of corn in the Midwest. Two species of rootworms that may cause severe damage to corn as both larvae and adults, the western and northern corn rootworms. Southern corn rootworm adults may damage corn leaves, however, because they cannot overwinter in most areas of the Midwest, southern corn rootworm larvae do not present a major threat to corn in this region.

Both sexes of western corn rootworm adults are yellow to green in color with a black stripe along the sides of their wing covers and are about 5/16 inch (7.5 mm) long. However, male wing covers are often nearly entirely black or at least darker in pigmentation than that of the female, which usually appears as more regular stripes. Slight variations in these color patterns may occur. Female western corn rootworm adults have larger abdomens than the males.

The adult of the northern corn rootworm is a tan to pale green beetle about 1/4 inch (6 mm) long. Newly emerged beetles are usually cream or light brown in color, but gradually turn green with age. No marked differences in coloration exist between sexes, but female beetles with their longer and larger abdomens are typically larger than the males.

The adult of the southern corn rootworm, also known as the spotted cucumber beetle, is about 3/8 inch (9 mm) long, yellow to green in color, with 11 conspicuous black spots on its back. Because adults and larvae of this species are considered of minor economic importance in the Midwest, the following discussion on damage, sampling, and management guidelines will not include the southern corn rootworm.

Male (left) and Female (right) western corn rootworms

Southern corn rootworm

It is extremely difficult to distinguish between larvae of the western and northern species. Rootworm larvae are white and slender, about 1/2 inch (13 mm) long when fully grown, have brown heads, and a dark plate on the top side of their "tails."

Larva feeding on corn root

Western and northern corn rootworm have only one generation per year. Eggs of both species are deposited in the soil by female beetles from mid-summer until autumn. The eggs overwinter and begin hatching from late May to early June in most areas of the Midwest. The newly hatched larvae seek out and feed on corn roots. The larvae pass through three stages, or instars, before pupating in the soil. Primary larval damage is caused by the later instars, typically from mid June through mid July.

By late June to early July, adult beetles begin to emerge from corn fields. The males emerge first, the females following in about 5 to 7 days. Mating takes place a short time thereafter. Within 2 weeks of emergence, females begin to lay eggs, usually toward the end of July, with peak egg laying occurring in early to mid August. The beetles in corn fields feed primarily on pollen, green silks, or leaves.

Several factors influence the development of a rootworm population in a particular field, and these should be considered when scouting. Rootworm beetles can readily move between fields and may cause damage in locations other than where they emerged. A late maturing corn field can attract large numbers of beetles since neighboring corn may have stopped producing pollen, the beetles' preferred food. Beetles may also move into corn and/or soybean fields which have an abundance of pollen-producing weeds, such as volunteer corn, ragweed, or foxtails. Some areas of the eastern Midwest have observed a steady and substantial increase of economic larval damage in first-year corn (corn following another crop). A behavioral variant of the western corn rootworm, that appears to have lost its fidelity for corn, can migrate to neighboring crops such as soybean to feed on pollen, flowers, and foliage, then lay eggs.

Where northern corn rootworm is the dominant species, a trait referred to as "extended diapause" can affect the survivability of the insect. Extended diapause describes a condition whereby a portion of northern corn rootworm eggs are capable of remaining dormant in the soil through two winters and one growing season before hatching in the second season. First-year corn in areas of Minnesota, Iowa, South Dakota, and

Wisconsin have been damaged by rootworms exhibiting this extended diapause trait. The western corn rootworm does not show this trait.

Northern corn rootworm

Soil texture is an important factor in regard to larval survival. Newly hatched rootworm larvae must move through the soil in search of corn roots and, as they move, they come in contact with soil particles. These particles, if coarse and abrasive, may scratch the cuticle of the larvae, resulting in death by desiccation. As a result, rootworms may be less of a problem in sandy soils. However, the effect of sandy soils can be modified by soil moisture. An irrigated or otherwise moist sandy soil will not affect rootworm populations as will the same soil under dry conditions. In addition, muck soils have shown a lower incidence of rootworm larval feeding damage and may provide some protection against this insect.

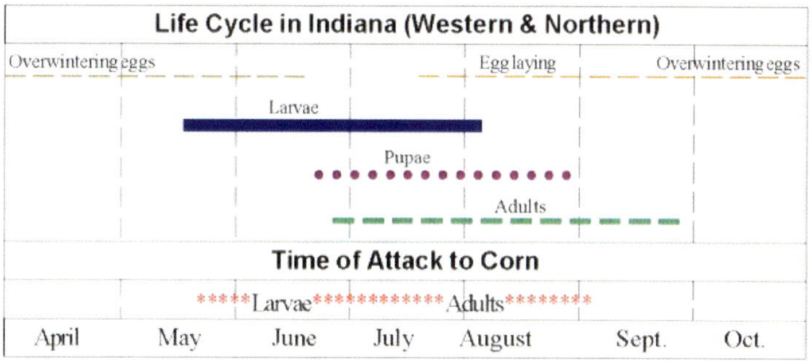

Damage

Western corn rootworm feeding

Both corn rootworm larvae and adults may damage corn plants. Newly hatched larvae feed primarily on root hairs and outer root tissue. As larvae grow and their food requirements increase, they burrow into the roots to feed. Larval damage is usually most severe after the secondary root system is well established and brace roots are developing. Root tips will appear brown and are often tunneled into and chewed back to the base of the plant. Larvae may be found tunneling into larger roots and occasionally in the plant crown.

Varying degrees of poor pollination from silk feeding

By reducing water and nutrient uptake of plants, larval root pruning places severe physiological stress on corn. Yield reductions may result, especially in corn also suffering from moisture, compaction, or fertility stress. Roots damaged by rootworms may be weakened to the extent that the plants lodge or grow in a curved "sled-runner" or "gooseneck" shape. These lodged and misshapen plants often pollinate poorly and are difficult to harvest, contributing to yield losses. Also, rootworm damage sites are often pathways for infection by pathogens. Thus, rootworm damage may be compounded by root and/or stalk rots.

High adult rootworm populations may interfere with corn plant pollination by severely clipping silks during pollen shed. This can result in poorly filled ears.

Corn rootworm damage

Western corn rootworm beetles can also cause damage by feeding on the green tissue of the leaves. Leaves may be partially or totally stripped by adults, resulting in a parchment-like appearance to the damaged area. Unless these beetles are unusually abundant, only scattered plants are severely damaged. However, such foliar damage can be indicative of a large adult population which may be a serious threat to pollination and next season's corn roots.

Though considered of minor importance in other crops such as soybean and alfalfa, the variant of the western corn rootworm can feed on pollen, flowers, and foliage in these other crops. The presence of beetles and the extent of damage to crops such as soybean may indicate the potential of economic damage to next year's corn roots, if rotating to corn.

Management Guidelines

For years, crop rotation has been a major management strategy for corn rootworm. As a general rule, rootworms cannot successfully complete their larval development on crops other than corn. Thus, if rootworm eggs are laid in a corn field one season and a soybean field the next season, the young larvae find themselves without a suitable host plant and soon starve to death in the soybean field. However, where the western corn rootworm variant or the northern corn rootworm extended diapause trait exists in the Midwest, crop rotation by itself is no longer an effective management strategy in all regions. Sampling for larvae and adults is imperative to apply the following management guidelines.

Larvae in Corn

1. Hand-Sorting of Soil:
    - Two or more rootworm larvae/plant prior to lay-by may signal the need for rootworm larval control.
2. Washing the Roots and Flotation:
    - Eight or more rootworm larvae/plant prior to lay-by may signal the need for rootworm larval control.

Adults in Corn

1. For Prevention of Silk Clipping and Interference with Pollination
2. Rootworm Larval Prediction for Next Year's Corn
3. Prediction of next year's larval problem, for those fields returning to corn, is based on the current year's average number of beetles per plant during late July through August.

Continuous Corn - Beetle counts in fields planted to corn for two or more years. Next season rotate to another crop or treat for rootworm larvae at cultivation or planting, if the average beetle count per plant equals or exceeds the number listed below for your average number of plants per acre.

Average Number Beetles/Plant		
Ave # Plants/Acre	"Normal" Soil Conditions	Severly Compacted Soils
14,000	1.4	1.0
16,000	1.3	0.9
18,000	1.1	0.8
20,000	1.0	0.7
22,000	0.9	0.6
24,000	0.8	0.6
26,000	0.8	0.5
28,000	0.7	0.5

Based on 1:1 ratio female to male

First-Year Corn - Beetle counts in first-year corn fields. Next season rotate to another crop or treat for rootworm larvae at cultivation or planting if the average beetle count per plant equals or exceeds the number listed below for your average number of plants per acre.

Average Number Beetles/Plant		
Ave # Plants/Acre	"Normal" Soil Conditions	Severly Compacted Soils
14,000	1.0	0.7
16,000	0.9	0.6
18,000	0.8	0.6
20,000	0.7	0.6
22,000	0.6	0.5
24,000	0.6	0.4
26,000	0.5	0.4
28,000	0.5	0.4

Based on 7:3 ratio Females to Males

Sequential Sampling - This sampling method, modified from the above, is based on sampling only enough beetles to determine the likelihood of an economic larval population developing in next year's corn, if going back to corn. Pairs of plants should be sampled as one walks a field in a U- shaped pattern. If fewer than 3 beetles are found on the first 10 plants, sampling may be terminated until the next week. If 3 to 17 beetles can be found on the first 10 plants, one should continue sampling. If more than 17 beetles, one may cease sampling since an economic population is present. If 54 plants have been sampled and counts are still in the intermediate range, stop sampling and return to the field and repeat the sampling procedure in one week. Weekly scouting should continue until an economic population is present or until the numbers of beetles decline in late August or early September.

Classifying Rootworm Beetle Populations Sampled Sequentially	
Plant Sampled	(continue sampling)
10	3-17
12	5-19
14	7-21
16	9-23
18	11-25
20	13-27
24	17-31
28	21-35
32	25-39
36	29-43
40	33-48
44	37-52
48	41-56
52	45-58
54	47-60

Adults in Soybean and Other Crops - Data is not available for the exact number of western corn rootworm beetles needed in soybean or other crops to cause economic larval damage in corn the following year. Consult with university researchers in the Midwest for new developments on the establishment of an economic threshold(s). Finding "numerous" western rootworm beetles versus a "few" in these crops during the sampling period may help one determine whether an economic problem in the following year's corn crop may occur.

Apple Maggot

Adult apple maggot fly, *Rhagoletis pomonella* (Walsh), lateral view.

The apple maggot, *Rhagoletis pomonella* (Walsh), also known as the apple maggot fly and railroad worm, is native to North America. Originally, it fed on the fruit of wild hawthorn (*Crataegus* spp.), but then became a primary pest of cultivated apples, especially in the northeastern United States and southeastern Canada. Varieties that mature in the summer and early fall are particularly vulnerable, but hard winter apples are sometimes infested. Thin-skinned, sweet, and subacid varieties are most susceptible, but acid varieties may be attacked.

Synonym

Trypeta pomonella Walsh

Distribution

Rhagoletis pomonella is a pest across eastern North America from eastern North Dakota and southern Manitoba to Nova Scotia, southward to eastern Texas to central Florida, occurring over the entire middle and eastern region of the United States. In 1981 it became established in Oregon and spread to California, Washington State, Utah and Colorado, and eventually Nebraska in 1991 (White and Elson-Harris 1994).

Description

Adults: The adult fly is about 4–5 mm, or slightly smaller than a house fly (6–7 mm). It is black in color, with white bands on the abdomen (four on the female and three on the male), and the wings are conspicuously marked with four oblique black bands.

Adult apple maggot fly, *Rhagoletis pomonella* (Walsh), dorsal view.

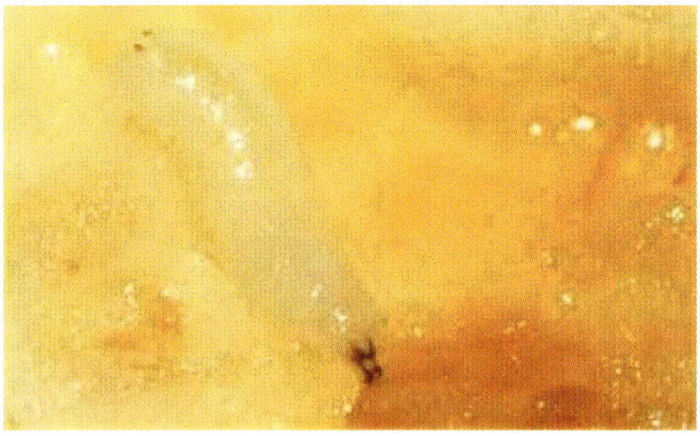

Larva of the apple maggot fly, *Rhagoletis pomonella* (Walsh).

Larvae: The larvae are white or yellowish tapered maggots slightly smaller than those of the house fly. The maggots are about 12 mm in length and tapering toward the head. Key characters for the separation of the larval stage from those of related species are given by Phillips (1946).

Life History and Habits

Adults emerge from the ground during early summer. Emergence continues for a month or more, and many pupae may remain inactive and not emerge until the second year. Egg-laying usually does not take place until eight to 10 days after the flies have emerged. The female punctures the skin of the fruit with her ovipositor and lays eggs singly in the pulp. Eggs hatch in five to 10 days. The maggots develop slowly in the green fruit and usually do not complete their growth until the infested fruits have dropped from the tree, after which growth is completed rapidly.

Larval development requires from two weeks in drops of early maturing apples to three or more months in hard winter varieties. Then the larvae leave the fruit and enter the soil where the puparia are formed. *Rhagoletis pomonella* undergoes a partial second generation in the southern part of the range, with adults emerging in early fall. Winter is passed as puparia in the soil.

Hosts

- apple and crab apple, *Malus* spp.
- cherry and plum, Prunus spp.
- Chickasaw plum, Prunus angustifolia
- chokeberry, Aronia arbutifolia
- cranberry, Vaccinium macrocarpum
- dogwood, Cornus florida
- hawthorn, Crataegus spp.
- peach, Prunus persica

Larvae have been found in pear, *Pyrus* spp., but no adults emerged. In New England, the fly uses the hips (fruits) of Japanese rose, *Rosa rugosa* and *Rosa carolina*, as alternative hosts. *Rhagoletis pomonella* has adapted to sour cherry, *Prunus cerasus*, in Utah, where it has not been reported from apple (White and Elson-Harris 1994).

While plum, pear and cherries also serve as hosts, usually the apple maggot is not a serious pest of these fruits. Crab apples are invariably infested by this pest. A closely related species, the blueberry maggot fly (*Rhagoletis mendax* Curran), is important as a pest of cultivated blueberries.

Damage

Injury to fruit is caused by maggots boring throughout the fruit, forming irregular, winding tunnels which turn brown, often causing premature dropping of fruit. When the fruit is slightly infested, there may be no external indication of the maggots, but when the fruit ripens, the burrows show as dark, winding trails beneath the skin. Minute egg punctures and distorted, pitted areas may show on the surface. Heavily infested early varieties of fruit will be reduced to a brown rotten masses filled with the fly larvae.

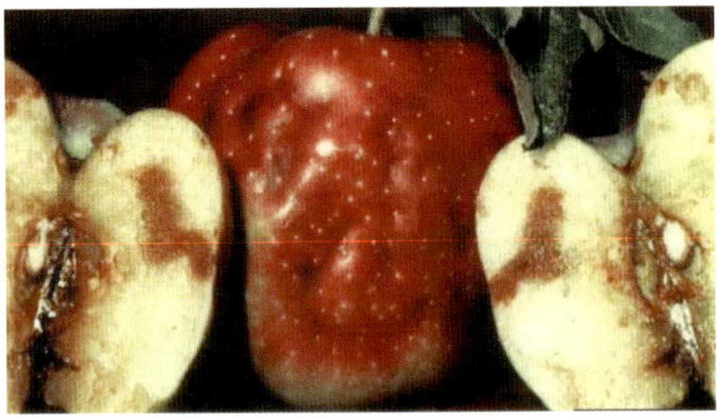

Damage caused by the apple maggot fly, *Rhagoletis pomonella* (Walsh).

Management

Cultural: The systematic destruction of infested apples and the elimination of hawthorn and abandoned apple trees in the vicinity of orchards are considered valid control practices. Apple maggots in fruits may be killed by placing the fruit in cold storage at 32°F for a period of 40 days.

Biological: Hymenopterous parasites recorded from *Rhagoletis pomonella* are a braconid wasp, *Opius melleus* Gahan, which attacks the larvae, and a tiny mymarid wasp, *Patasson conotracheli* (Girault), which attacks the eggs. Because the apple maggot feeds within fruit, biological control agents have not been very effective (Varela 2008).

Chemical: Emergence and dispersal of adult flies must be carefully monitored to effectively time treatments. Sticky traps, including yellow rectangles and red spheres, are used in other areas to monitor adults and time treatments. Unfortunately, only provisional economic thresholds are available for apple maggots, even in areas where it has long been a pest. You can detect the first emergence of adults by hanging yellow sticky traps in abandoned orchards or unsprayed apple trees in infested areas. To detect the beginning of egg laying, hang red sticky spheres in apple trees, then treat as soon as the first fly is found. In Oregon, where some orchards are treated regularly for apple maggots, the first maggot spray is applied seven to 10 days after the first fly has emerged. Later sprays follow at 10 to 14 day intervals as long as adults are active and are being caught in traps (Varela 2008).

Beet Armyworm

The beet armyworm originated in Southeast Asia. It was first discovered in North America about 1876, when it was found in Oregon, and it reached Florida in 1924. It rarely overwinters in areas where frost kills its host plants. Thus, overwintering is generally limited to Arizona, Florida, and Texas. Nevertheless, beet armyworm invades the southern half of the United States (Maryland to Colorado to northern California, and south) annually. Except in greenhouses, it rarely is a pest except in southern states.

Description and Life Cycle

Seasonal activity varies considerably according to climate. In warm locations such as Florida, all stages can be found throughout the year, although development rate and overall abundance are reduced during the winter months (Tingle and Mitchell 1977). The life cycle can be completed in as few as 24 days, and six generations have been reared during five months of summer weather in Florida (Wilson 1934).

Newly hatched larvae of the beet armyworm, *Spodoptera exigua* (Hübner).

Egg: Eggs are laid in clusters of 50 to 150 eggs per mass. Normal egg production is about 300 to 600 per female. Eggs are usually deposited on the lower surface of the leaf, and often near blossoms and the tip of the branch. The individual eggs are circular when viewed from above, but when examined from the side the egg is slightly peaked, tapering to a point. The eggs are greenish to white in color, and covered with a layer of whitish scales that gives the egg mass a fuzzy or cottony appearance. Eggs hatch in two to three days during warm weather.

Larva: There normally are five instars, although additional instars are sometimes reported. Duration of the instars under warm (summer) conditions is reported to be 2.3, 2.2, 1.8, 1.0, and 3.1 days, respectively (Wilson 1932), and at constant 30°C instar development time was reported by Fye and McAda (1972) to be 2.5, 1.5, 1.2, 1.5, and 3.0 days, respectively. Head capsule widths average 0.25, 0.45, 0.70, 1.12, and 1.80 mm, respectively.

The larvae are pale green or yellow in color during the first and second instars, but acquire pale stripes during the third instar. During the fourth instar, larvae are darker dorsally, and possess a dark lateral stripe. Larvae during the fifth instar are quite variable in appearance, tending to be green dorsally with pink or yellow color ventrally and a white stripe laterally. A series of dark spots or dashes is often present dorsally and dorsolaterally. Sometimes larvae are very dark in color, even black. The spiracles are white with a narrow black border. The body is practically devoid of hairs and spines. In the southern states, the larva of beet armyworm is easily confused with southern armyworm, *Spodoptera eridania* (Cramer), but southern armyworm can be distinguished by the presence of a large dark spot laterally on the first abdominal segment that disrupts the lateral stripe. Beet armyworm occasionally bears a spot laterally, but if present it occurs on the mesothorax, not on the first abdominal segment.

Pupa: Pupation occurs in the soil. The chamber is constructed from sand and soil particles held together with an oral secretion that hardens when it dries. The pupa is light brown in color and measures about 15 to 20 mm in length. Duration of the pupal stage is six to seven days during warm weather.

Partly grown larva of the beet armyworm, *Spodoptera exigua* (Hübner).

Adult: The moths are moderately sized, the wing span measuring 25 to 30 mm. The forewings are mottled gray and brown, and normally with an irregular banding pattern and a light colored bean-shaped spot. The hind wings are a more uniform gray or white color, and trimmed with a dark line at the margin. Mating occurs soon after emergence of the moths, and oviposition begins within two to three days. Oviposition extends over a three to seven day period, and the moths usually perish within nine to 10 days of emergence. Heppner (1998) provides a key to the adults of North American *Spodoptera*.

Mature larva of beet armyworm, *Spodoptera exigua* (Hübner).

Adult beet armyworm, *Spodoptera exigua* (Hübner).

Host Plants

The beet armyworm has a wide host range, occurring as a serious pest of vegetable, field, and flower crops. Among susceptible vegetable crops are asparagus, bean, beet, broccoli, cabbage, cauliflower, celery, chickpea, corn, cowpea, eggplant, lettuce, onion, pea, pepper, potato, radish, spinach, sweet potato, tomato, and turnip. Field crops damaged include alfalfa, corn, cotton, peanut, safflower, sorghum, soybean, sugarbeet, and tobacco. Weeds also are suitable for larval development, including such common plants as lambsquarters, *Chenopodium album*; mullein, *Verbascum* sp.; pigweed,

Amaranthus spp.; purslane, *Portulaca* spp.; Russian thistle, *Salsola kali*; parthenium, *Parthenium* sp.; and tidestromia, *Tidestromia* sp.

As with many polyphagous insects, although the host range is potentially very broad, the insect has definite food preferences, and the suitabilities of the food plants vary as well. Greenberg et al. 2001 compared growth of beet armyworm larvae on three crops (cabbage, cotton, pepper) and two weeds (redroot pigweed, sunflower). They reported that the level of relative consumption was cabbage, followed by pepper, sunflower, cotton, and pigweed. Relative growth, judged by pupal weights, was greatest on pigweed, equal for cotton, cabbage, and pepper, and least on sunflower. The larvae also matured fastest on pigweed. The authors interpreted these results to indicate that pigweed was nutritionally superior, and that cabbage was the poorest host.

Damage

Larvae feed on both foliage and fruit. In Florida it is regarded as a serious defoliator of flower crops and cotton, though much of the injury is induced by insecticide use that interferes with natural enemy activity. Young larvae feed gregariously and skeletonize foliage. As they mature, larvae become solitary and eat large irregular holes in foliage. They also burrow into the crown or center of the head on lettuce, or on the buds of cole crops. As a leaf feeder, beet armyworm consumes much more cabbage tissue than the diamondback moth, *Plutella xylostella* (Linnaeus), but is less damaging than the cabbage looper, *Trichoplusia ni* (Hübner) (East et al. 1989), as the latter grows to be a much larger caterpillar.

Tomato fruit is quite susceptible to injury, especially near fruit maturity, but beet armyworm is not considered to be as threatening to tomato as is the corn earworm, *Helicoverpa zea* (Boddie) (Zalom et al. 1986). In tomato crops, infestation early in the growing period is more damaging than later infestation, and as little as 1 caterpillar per 20 tomato plants can cause economic loss (Taylor and Riley 2008)

Sampling

Pheromone traps can be used to detect the presence of adult beet armyworm. Visual sampling for damage and larvae, combined with an action threshold of 0.3 larvae per plant, was used successfully on cabbage in south Texas to determine the need for crop treatment with insecticides (Cartwright et al. 1987). Regular monitoring of crops, probably about twice per week, is recommended because adults frequently invade from surrounding crops or weeds.

Natural Enemies

Numerous native natural enemies have adapted to this pest. Among the most common parasitoids are *Chelonus insularis* Cresson, *Cotesia marginiventris* (Cresson), and *Meteorus autographae* (Muesbeck) (all Hymenoptera: Braconidae), and the tachinid

Lespesia archippivora (Riley) (Diptera: Tachinidae) (Oatman and Platner 1972, Ruberson et al. 1994). Predators frequently attack the eggs and small larvae; among the most important are minute pirate bugs, *Orius* spp. (Hemiptera: Anthocoridae); bigeye bugs, *Geocoris* spp. (Hemiptera: Geocoridae); damsel bugs, *Nabis* spp. (Hemiptera: Nabidae); and a predatory stink bug, *Podisus maculiventris* (Say). Pupae are subject to attack, especially by the red imported fire ant, *Solenopsis invicta* Buren. Fungal diseases, *Erynia* sp. and *Nomurea rileyi*, and a nuclear polyhedrosis virus also inflict some mortality (Wilson 1933, 1934; Ruberson et al. 1994). The important mortality factors vary among crops, and among geographic regions. None except the nuclear polyhedrosis virus are highly specific to beet armyworm, which may explain why they are not especially effective. Virus is considered to be the most important mortality factor.

Insecticides

In the Southeast and Southwest, the relatively high abundance of beet armyworm has stimulated frequent application of insecticides to foliage. Insecticide resistance is a major problem in management of this insect, possibly because it attacks crops such as flowers, cotton, and vegetables - crops that are treated frequently with insecticides. Beet armyworm abundance is favored by frequent insecticide use, and it is considered to be a secondary or induced pest in some crops (Eveleens et al. 1973). Also, intensive use of insecticides for beet armyworm control in vegetables such as celery has stimulated outbreaks of other pests, principally American serpentine leafminer, *Liriomyza trifolii* (Burgess).

Beet armyworm larvae are susceptible to neem products (Prabhaker et al. 1986). Eggs can be killed with petroleum oil, and both eggs and young larvae can be controlled with foliar applications of 5% cottonseed oil, but this concentration is damaging to some plants (Butler and Henneberry 1990).

Pheromones can also be used to disrupt mating and inhibit or eliminate reproduction. Saturation of the atmosphere around beet armyworm-susceptible crops has been estimated to reduce mating by 97% (Wakamura and Takai 1992).

Biological Control

Several insect pathogens may prove to be useful for suppression of beet armyworm. A nuclear polyhedrosis virus isolated from beet armyworm is fairly effective as a bioinsecticide under greenhouse conditions, where inactivation by ultraviolet light in sunlight is not a severe problem. The fungus *Beauveria bassiana* has the same attributes and limitations. Entomopathogenic nematodes (Rhabditida: Steinernematidae and Heterorhabditidae) successfully infect both larvae and adults of beet armyworm, though it is difficult to attain suppression of foliage-feeding insects under field conditions because the nematodes are favored by moisture. Thus, the nematodes are more effective at killing the larvae when they drop to the soil to pupate; the prepupal stage is more susceptible than the pupal stage.

Aphids

This native ladybug species is the best known garden predator available

There are approximately 4,000 aphid species found throughout the world. Low to moderate numbers are usually not harmful to plants and rarely require control. However, heavy infestations will cause leaves to curl, wilt or yellow and stunted plant growth. A general decline in overall plant vigor will also be noticed. Several species can transmit plant diseases, particularly viruses which they pass on during feeding.

Identification

Aphids are small (1/8 inch long), soft bodied, pear-shaped insects that may be green, yellow, brown, red or black in color depending on species and food source. Generally adults are wingless, but some can grow wings, especially if populations are high. They have two whip-like antennae at the tip of the head and a pair of tube-like structures, called cornicles, projecting backward out of their hind end.

As they feed, aphids secrete large amounts of a sticky fluid known as honeydew. This sweet goo drips onto plants, attracting ants and promoting a black sooty mold growth on leaves. Cars and lawn furniture that are under infested trees will also be covered with this sticky fluid.

Life Cycle

In spring wingless female aphids hatch from overwintering eggs and soon give birth to many nymphs (males are not present). Young nymphs increase gradually in size and within a week give birth to many more nymphs. This process is repeated several times and results in huge population explosions. As the colony grows, a few of the females develop wings and fly off to other host plants to start new colonies. In late summer and early fall sexual forms (males and females) develop which mate and lay overwintering eggs. There are many overlapping generations per year.

Most aphids, except for the sexual forms, do not have to mate in order to reproduce, and they produce live young, rather than eggs.

Adult and nymph aphids.

A wide range of aphid species can affect field crops. Most are small (only a couple of millimetres long) with oval-shaped green, brown or black bodies. Often occurring in colonies, aphids suck on sap, causing loss of vigour, and in some cases yellowing, stunting or distortion of plant parts. Honeydew (unused sap) secreted by the insects can cause sooty mould to develop on leaves. In crops such as cotton, the honeydew affects fibre quality. Aphids can also be vectors (carriers) for viruses.

Their feeding rarely causes major damage in most crops, and control measures are usually not warranted, as a range of parasites and predators keep population numbers down. Exceptions may be where the crop is under moisture stress, or where heavy populations are observed (high populations aggravate the effect of moisture stress). For example, between 30 to 50 aphids per plant could be used as a threshold level for wheat.

Do Aphids Have a Benefit?

In many cases, "pests" have a positive side, but are there any benefits aphids provide? Aphids don't help the plant itself, however, beneficial bugs like lady beetles and lacewings love feasting on aphids. Since lady beetles can eat many aphids per day, they can be attracted to and stay in your garden to help you keep the aphid population in check.

How to Identify Aphid Damage

Keeping a healthy organic garden can become a major issue if you have some of these pesky insects running around.

Aphids, also known as Plant Lice or Green Flies, infest both gardens and landscapes. Effective aphid control is essential for the overall health of your organic garden.

If certain plant roots are attacked by aphids, the plant may shrivel and die, particularly those of young plants or new growth.

Aphids prefer the bottoms of leaves rather than the top, and they are so small they can be difficult to see at a glance. It's important to keep a watchful eye on your plants for signs of aphid damage, since the aphids themselves may not be visible.

Visible Symptoms of Aphid Damage

Even if you can't see the aphids, they leave behind plenty of signs. Keep an eye out for some of the following indicators:

- Aphids produce a sweet, sticky excretion known as honeydew, and if you notice a substance on your plants that feels sticky, it may mean aphids are living on the plant.
- The honeydew secretion may turn black if it is affected by sooty mold fungus. The black spots that the fungus creates on the leaves will prevent the plant from absorbing the light it needs for photosynthesis. Without the ability to photosynthesize, the plant will eventually die. Sooty mold fungus thrives from the honeydew secretions of the aphids because the honeydew contains necessary ingredients like amino acids, protein, and sugar the fungus requires. If aphids have decided to call your plants home, it's also a prime environment for this fungus to thrive.
- Leaves may curl and turn yellowish if there is a large infestation of aphids on a plant or tree. This is the time you will want to use a product for effective aphid control before they do more damage to your plant and garden.
- There may be an increase in the ant population surrounding an infested plant. Since the aphids' honeydew is high in sugar, it attracts the ants to the plant. The ants feed off the honeydew secretions and actually protect the aphids from predators.

THERE MAY BE AN INCREASE IN THE ANT POPULATION SURROUNDING AN INFESTED PLANT

Results of an Aphid Infestation

There are many different species of aphids, and some have affinities for specific types of plants. Regardless of which species is causing your aphid infestation, these pesky bugs can be detrimental to your plants. Here are some of the common results of an aphid infestation:

- When aphids devour the liquids from the plants, they also become carriers of any viruses the host plant may have. When the aphid moves to a different plant to begin a new colony, these viruses can be transmitted from one plant to another by the aphid. It only takes one aphid to transmit a virus to a plant, and the viruses can even be more detrimental than the aphids themselves.
- Plants may stop growing or have yellowing leaves. This can occur on any plant such as tomatoes, pumpkins, zucchini, beans, and more. Once the plant stops growing or its leaves begin to yellow, its yield won't be as high. Like the name suggests, the rose aphid is particularly attracted to roses. Without aphid control on roses, they reproduce at a rapid rate and can cause bloom deformities as well as a reduction in the number and quality of blooms. Tomato plants are attractive to the potato and green peach aphids. While the plants can withstand a relatively high number of aphids before beginning to show symptoms, getting rid of aphids will protect the plant from withering leaves and a reduced tomato crop.
- If certain plant roots are attacked by aphids, the plant may shrivel and die, particularly those of young plants or new growth.

APHIDS TEND TO COLONIZE ON NEW GROWTH AND FLOWER BUDS

- Aphids deplete the plant's fluids and nutrients from all parts, including leaves, stems, new growth, and flowers, so plant growth may also be impacted.
- Aphids tend to colonize on new growth and flower buds, so, flowers, buds, or new growth may appear stunted, distorted, or otherwise unhealthy.

- Larger plants and trees may be able to survive the onslaught of aphids with some damage, but they may not be quite as strong or productive as they were prior to the aphid infestation.
- The presence of honeydew may encourage the growth of fungi, such as sooty mold fungus and others which can cause further damage to your plants.

Caterpillars

The caterpillar is the larvae (the baby) of both a butterflyand a moth. After around 2-3 weeks, the caterpillar builds itself into a cocoon where it remains a pupa for a further 2 weeks. The caterpillar then emerges having grown wings.

The moth caterpillar is well known for being a pest, particularly in the fabric industry. One species of caterpillar has destroyed reams of silk in the far east, known in China as a silkworm.

Generally, most species of caterpillar are considered to be agricultural pests as they can munch their ways through fields of crops, often leaving enormous holes which result in unhealthy or inedible plants.

Some species of caterpillar are also highly poisonous, particularly those species that live in the tropical rain forests. Other species are only poisonous in their caterpillar form, meaning when they turn into a moth or butterfly, they no longer have venom.

There are more than 20,000 different species of known caterpillar found all around the world and it is estimated that there are many more that are undiscovered as new species of butterflies and moths are regularly found in regions where there is little, if any, human presence.

Caterpillars difference in size, colour and appearance depending on their species. Some caterpillars are very brightly coloured where other caterpillar species are quite dull-looking in comparison. Some species of caterpillars are very hairy, where others are very smooth. The main aim of the appearance of the caterpillar is to intimidate its predators and to deter them from eating it.

The caterpillar, like the butterfly, is a herbivorous animal but the diets of the caterpillar and the butterfly are very different. Butterflies use their long straw-like tongues to drink the nectar out of flowers, which is an adaptation that occurs in the process where the caterpillar becomes a butterfly. Caterpillars mainly feed on leaves, plants and flowering plants and big holes can often be found in leaves indicating the presence of a caterpillar.

A number of species of caterpillar are known to be carnivorous and eat a variety of insects that pass them. One caterpillar species found in Hawaii, hooks itself onto a leaf where it stands very straight trying to catching passing bugs.

Due to their small size and worm-like shape, caterpillars are preyed upon by numerous different species of animal but the main predators of the caterpillar are birds and large insects such as wasps. Caterpillars are also commonly preyed upon by small mammals and reptiles.

Description

Crochets on a Caterpillar's Prolegs

Larvae of *Craesus septentrionalis*, a sawfly showing 6 pairs of pro-legs.

Caterpillars have soft bodies that can grow rapidly between moults. Their size varies between species and instars (moults) from as small as 1 mm up to 14 cm. Only the head capsule is hardened. The mandibles are tough and sharp for chewing (this contrasts with most adult Lepidoptera, which have highly reduced or soft mandibles). Behind the mandibles of the caterpillar are the spinnerets, for manipulating silk. Antennae are present on either side of the labrum, but small and relatively inconspicuous.

Some larvae of the order Hymenoptera (ants, bees and wasps) can appear like the caterpillars of the Lepidoptera. Such larvae are mainly seen in the sawfly suborder. However while these larvae superficially resemble caterpillars, they can be distinguished by the presence of prolegs on every abdominal segment, an absence of crochets or hooks on the prolegs (these are present on lepidopteran caterpillars), one pair of prominent ocelli on the head capsule, and an absence of the upside-down Y-shaped suture on the front of the head.

Lepidopteran caterpillars can be differentiated from sawfly larvae by:

- the numbers of pairs of pro-legs; sawfly larvae have 6 or more pairs while caterpillars have a maximum of 5 pairs.
- the number of stemmata (simple eyes); the sawfly larvae have only two, while caterpillars usually have six.
- the presence of crochets on the prolegs; these are absent in the sawflies.
- sawfly larvae have an invariably smooth head capsule with no cleavage lines, while lepidopterous caterpillars bear an inverted "Y" or "V" (adfrontal suture).

Defenses

The saddleback caterpillar has urticating hair and aposematic colouring.

Many animals feed on caterpillars as they are rich in protein. As a result, caterpillars have evolved various means of defense.

Caterpillars have evolved defenses against physical conditions such as cold, hot or dry environmental conditions. Some Arctic species like *Gynaephora groenlandica* have special basking and aggregation behaviours apart from physiological adaptations to remain in a dormant state.

Appearance

Costa Rican hairy caterpillar. The spiny bristles are a self-defense mechanism

The appearance of a caterpillar can often repel a predator: its markings and certain body parts can make it seem poisonous, or bigger in size and thus threatening, or non-edible. Some types of caterpillars are indeed poisonous or distasteful and their bright coloring is aposematic. Others may mimic dangerous caterpillars or other animals while not being dangerous themselves. Many caterpillars are cryptically colored and resemble the plants on which they feed. An example of caterpillars that use camouflage for defence is the species *Nemoria arizonaria*. If the caterpillars hatch in the spring and feed on oak catkins they appear green. If they hatch in the summer they appear dark colored, like oak twigs. The differential development is linked to the tannin content in the diet. Caterpillars may even have spines or growths that resemble plant parts such as thorns. Some look like objects in the environment such as bird droppings.

Chemical Defenses

More aggressive self-defense measures are taken by some caterpillars. These measures include having spiny bristles or long fine hair-like setae with detachable tips that will irritate by lodging in the skin or mucous membranes. However some birds (such as cuckoos) will swallow even the hairiest of caterpillars. Other caterpillars acquire toxins from their host plants that render them unpalatable to most of their predators. For instance, ornate moth caterpillars utilize pyrrolizidine alkaloids that they obtain from their food plants to deter predators. The most aggressive caterpillar defenses are bristles associated with venom glands. These bristles are called urticating hairs. A venom which is among the most potent defensive chemicals in any animal is produced by the South American silk moth genus *Lonomia*. Its venom is an anticoagulant powerful enough to cause a human to hemorrhage to death. This chemical is being investigated for potential medical applications. Most urticating hairs range in effect from mild irritation to dermatitis.

Giant swallowtail caterpillar everting its osmeterium in defense

Plants contain toxins which protect them from herbivores, but some caterpillars have evolved countermeasures which enable them to eat the leaves of such toxic plants. In addition to being unaffected by the poison, the caterpillars sequester it in their body, making them highly toxic to predators. The chemicals are also carried on into the adult stages. These toxic species, such as the cinnabar moth (*Tyria jacobaeae*) and monarch (*Danaus plexippus*) caterpillars, usually advertise themselves with the danger colors of red, yellow and black, often in bright stripes. Any predator that attempts to eat a caterpillar with an aggressive defense mechanism will learn and avoid future attempts.

Some caterpillars regurgitate acidic digestive juices at attacking enemies. Many papilionid larvae produce bad smells from extrudable glands called osmeteria.

Defensive Behaviors

Caterpillars linked together into a "train"

Many caterpillars display feeding behaviors which allow the caterpillar to remain hidden from potential predators. Many feed in protected environments, such as enclosed inside silk galleries, rolled leaves or by mining between the leaf surfaces.

Some caterpillars, like early instars of the tomato hornworm and tobacco hornworm, have long "whip-like" organs attached to the ends of their body. The caterpillar wiggles these organs to frighten away flies and predatory wasps. Some caterpillars can evade predators by using a silk line and dropping off from branches when disturbed. Many species thrash about violently when disturbed to scare away potential predators. One species (*Amorpha juglandis*) even makes high pitched whistles that can scare away birds.

Social Behaviors and Relationships with other Insects

Some caterpillars obtain protection by associating themselves with ants. The Lycaenid butterflies are particularly well known for this. They communicate with their ant protectors by vibrations as well as chemical means and typically provide food rewards.

Some caterpillars are gregarious; large aggregations are believed to help in reducing the levels of parasitization and predation. Clusters amplify the signal of aposematic coloration, and individuals may participate in group regurgitation or displays.

Pine processionary (*Thaumetopoea pityocampa*) caterpillars often link into a long train to move through trees and over the ground. The head of the lead caterpillar is visible, but the other heads can appear hidden.

Predators

Caterpillars suffer predation from many animals. The European pied flycatcher is one species that preys upon caterpillars. The flycatcher typically finds caterpillars among oak foliage. Paper wasps, including those in the genus *Polistes* and *Polybia* catch caterpillars to feed their young and themselves.

Behavior

Caterpillars have been called "eating machines", and eat leaves voraciously. Most species shed their skin four or five times as their bodies grow, and they eventually enter a pupal stage before becoming adults. Caterpillars grow very quickly; for instance, a tobacco hornworm will increase its weight ten-thousandfold in less than twenty days. An adaptation that enables them to eat so much is a mechanism in a specialized midgut that quickly transports ions to the lumen (midgut cavity), to keep the potassium level higher in the midgut cavity than in the hemolymph.

A pasture day moth caterpillar feeding on capeweed

A gypsy moth caterpillar

Most caterpillars are solely herbivorous. Many are restricted to feeding on one species of plant, while others are polyphagous. Some, including the clothes moth, feed on detritus. Some are predatory, and may prey on other species of caterpillars (e.g. Hawaiian *Eupithecia*). Others feed on eggs of other insects, aphids, scale insects, or ant larvae. A few are parasitic on cicadas or leaf hoppers (Epipyropidae). Some Hawaiian caterpillars (*Hyposmocoma molluscivora*) use silk traps to capture snails.

Many caterpillars are nocturnal. For example, the "cutworms" (of the family Noctuidae) hide at the base of plants during the day and only feed at night. Others, such as gypsy moth (Lymantria dispar) larvae, change their activity patterns depending on density and larval stage, with more diurnal feeding in early instars and high densities.

Economic Effects

Caterpillars cause much damage, mainly by eating leaves. The propensity for damage is enhanced by monocultural farming practices, especially where the caterpillar is specifically adapted to the host plant under cultivation. The cotton bollworm causes enormous losses. Other species eat food crops. Caterpillars have been the target of pest control through the use of pesticides, biological control and agronomic practices. Many species have become resistant to pesticides. Bacterial toxins such as those from *Bacillus thuringiensis* which are evolved to affect the gut of Lepidoptera have been used in sprays of bacterial spores, toxin extracts and also by incorporating genes to produce them within the host plants. These approaches are defeated over time by the evolution of resistance mechanisms in the insects

Hypsipyla grandela damages mahogany in Brazil

Plants evolve mechanisms of resistance to being eaten by caterpillars, including the evolution of chemical toxins and physical barriers such as hairs. Incorporating host plant resistance (HPR) through plant breeding is another approach used in reducing the impact of caterpillars on crop plants.

Some caterpillars are used in industry. The silk industry is based on the silkworm caterpillar.

Human Health

Caterpillar hair can be a cause of human health problems. Caterpillar hairs sometimes have venoms in them and species from approximately 12 families of moths or butterflies worldwide can inflict serious human injuries ranging from urticarial dermatitis and atopic asthma to osteochondritis, consumption coagulopathy, renal failure, and intracerebralhemorrhage. Skin rashes are the most common, but there have been fatalities. *Lonomia* is a frequent cause of envenomation in Brazil, with 354 cases reported between 1989 and 2005. Lethality ranging up to 20% with death caused most often by intracranial hemorrhage.

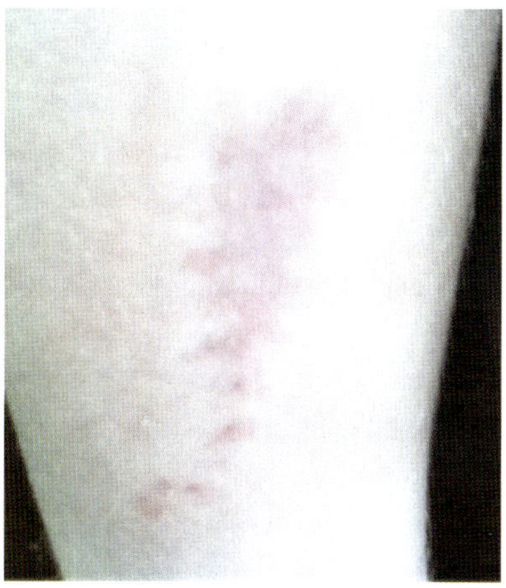

Buck moth caterpillar sting on a shin twenty-four hours after occurrence in south Louisiana. The reddish mark covers an area about 20 mm at its widest point by about 70 mm in length.

Caterpillar hair have also been known to cause kerato-conjunctivitis. The sharp barbs on the end of caterpillar hairs can get lodged in soft tissues and mucous membranes such as the eyes. Once they enter such tissues, they can be difficult to extract, often exacerbating the problem as they migrate across the membrane.

This becomes a particular problem in an indoor setting. The hair easily enter buildings through ventilation systems and accumulate in indoor environments because of their small size, which makes it difficult for them to be vented out. This accumulation increases the risk of human contact in indoor environments.

Caterpillars are a food source in some cultures. For example, in South Africa mopane worms are eaten by the bushmen, and in China silkworms are considered a delicacy.

A Serious Pest

Caterpillars, with their cutting and chewing mouthparts, or mandibles, as was written before, can cause an exceptional amount of damage if not treated early-on or prevented. The key to coping with caterpillars does not necessarily lie in the thoroughness of your scouting but more in the knowledge of your foe and methodology and timing of your effort. Your area university's cooperative extension office — since caterpillars are usually regional in distribution — will probably be able to provide information as to what species are known to be problematic in your area and at what time of year they are doing what they do. The extension agent may also be able to prescribe curative action, but may not be well-versed in the Green Methods, and thus may not be able to offer fully all the nontoxic ways to get the job done — and there are many.

Moths and butterflies, as we've indicated, develop through a complete metamorphosis: egg, larva, pupa (chrysalis) and adult, which is usually done at a specific time of year based on species. (Unlike aphids which develop constantly throughout the season by way of an incomplete metamorphosis, molting through multiple immature nymphal stage before reaching, uninterrupted by pupation, their adulthood.) In light of this, scouting will be done differently at various times during the season depending on what species is being scouted. Moreover, if you see a caterpillar with "eggs" on its back, they're not eggs as that's not how these things work. More than likely the "eggs" you're seeing are actually the pupae of parasitic wasps in cocoons of their own.

Grasshoppers

The grasshopper is a medium to large sized insect and the grasshopper is found (close to grass) all over the world. Grasshoppers are best known for their ability to jump incredible heights and distances. Most grasshopper individuals grow to about 2 inches long although larger grasshoppers are found on a fairly regular basis that grow to more than 5 inches in length. The grasshopper has wings meaning it can migrate over long distances when the weather gets too cold. There are 11,000 thousand known species of grasshopper on Earth, that live in grassy areas such as fields and meadows and forest and woodland. Like all insects, all species of grasshopper have a three-part body that is made up of the grasshopper's head, it's thorax and the abdomen. Grasshoppers also have six legs, two pairs of wings, and two antennae. The antennae of the grasshopper are known to be remarkably long and can often be longer than the grasshopper's body, although the grasshopper's antennae and the grasshopper's body are normally about the same size. Grasshoppers use their long antennae in order to make sense of their surroundings. Grasshoppers have six jointed legs that are incredibly powerful for such a small creature, as grasshoppers are able to jump extraordinary distances. The two back legs of the grasshopper are long and powerful and are just for jumping, where the four front legs of the grasshopper are primarily used to hold onto prey and to help it to walk. Despite their large size, grasshoppers are herbivores animals and have a diet that consists solely of plant matter. Grasshoppers eat grasses, weeds, leaves, shrubs, bark and numerous other species of plants that surround them. The grasshopper is also a stable food source for many predators around the world including reptiles, insects, small mammals and birds. It is common for humans to eat grasshoppers in places like Asia and Africa where the bigger species of grasshopper are found, and there is a less readily available alternative protein source. The female grasshopper lays an egg pod that contains a couple of dozen grasshopper eggs in the late autumn to early winter depending on the area. The female grasshopper inserts her egg pod into the soil so that it is a couple of inches underground. The grasshopper eggs can take up to 9 months to hatch as they wait until the weather has warmed before breaking into the outside world. When the first baby grasshopper (known as a nymph) hatches out of its egg, it tunnels through the soil and up to the surface, and the remaining grasshopper nymph follow. As they get older, the grasshoppers will increase in size until they are adults. The grasshopper only remains in this stage (young and adult) for a few months before it dies meaning that most grasshopper individuals spend the majority of their lives inside an egg.

Predators, Parasites and Pathogens

Cottontop tamarin monkey eating a grasshopper

Grasshopper with parasitic mites

Grasshoppers have a wide range of predators at different stages of their lives; eggs are eaten by bee-flies, ground beetles and blister beetles; hoppers and adults are taken by other insects such as ants, robber flies and sphecid wasps, by spiders, and by many birds and small mammals.

The eggs and nymphs are under attack by parasitoids including blow flies, flesh flies, and tachinid flies. External parasites of adults and nymphs include mites.Female grasshoppers parasitised by mites produce fewer eggs and thus have fewer offspring than unaffected individuals.

The grasshopper nematode (*Mermis nigrescens*) is a long slender worm that infects grasshoppers, living in the insect's hemocoel. Adult worms lay eggs on plants and the host becomes infected when the foliage is eaten. *Spinochordodes tellinii* and *Paragordius tricuspidatus* are parasitic worms that infect grasshoppers and alter the behaviour of their hosts. When the worms are sufficiently developed, the grasshopper is persuaded to leap into a nearby body of water where it drowns, thus enabling the parasite to continue with the next stage of its life cycle, which takes place in water.

Grasshoppers are affected by diseases caused by bacteria, viruses, fungi and protozoa. The bacteria *Serratia marcescens* and *Pseudomonas aeruginosa* have both been implicated in causing disease in grasshoppers, as has the entomopathogenic fungus*Beauveria bassiana*. This widespread fungus has been used to control various pest insects around the world, but although it infects grasshoppers, the infection is not usually lethal because basking in the sun has the result of raising the insect's temperature above a threshold tolerated by the fungus. The fungal pathogen *Entomophaga grylli* is able to influence the behaviour of its grasshopper host, causing it to climb to the top of a plant and cling to the stem as it dies. This ensures wide dispersal of the fungal spores liberated from the corpse.

Locusts killed by the naturally occurring fungus *Metarhizium*

The fungal pathogen *Metarhizium acridum* is found in Africa, Australia and Brazil where it has caused epizootics in grasshoppers. It is being investigated for possible use as a microbial insecticide for locust control. The microsporidian fungus *Nosema locustae*, once considered to be a protozoan, can be lethal to grasshoppers. It has to be consumed by mouth and is the basis for a bait-based commercial microbial pesticide. Various other microsporidians and protozoans are found in the gut.

Anti-predator Defences

Grasshoppers exemplify a range of anti-predator adaptations, enabling them to avoid detection, to escape if detected, and in some cases to avoid being eaten if captured. Grasshoppers are often camouflaged to avoid detection by predators that hunt by sight; some species can change their coloration to suit their surroundings.

Several species such as the hooded leaf grasshopper *Phyllochoreia ramakrishnai* (Eumastacoidea) are detailed mimics of leaves. Stick grasshoppers (Proscopiidae) mimic wooden sticks in form and colouration. Grasshoppers often have deimatic patterns on their wings, giving a sudden flash of bright colours that may startle predators long enough to give time to escape in a combination of jump and flight.

Some species are genuinely aposematic, having both bright warning coloration and sufficient toxicity to dissuade predators. *Dictyophorus productus* (Pyrgomorphidae) is a "heavy, bloated, sluggish insect" that makes no attempt to hide; it has a bright red abdomen. A *Cercopithecus* monkey that ate other grasshoppers refused to eat the species. Another species, the rainbow or painted grasshopper of Arizona, *Dactylotum bicolor* (Acridoidea), has been shown by experiment with a natural predator, the little striped whiptail lizard, to be aposematic.

Gaudy grasshopper, *Atractomorpha lata*, evades predators with camouflage.

Leaf grasshopper, *Phyllochoreia ramakrishnai*, mimics a green leaf.

Painted grasshopper, *Dactylotum bicolor*, deters predators with warning coloration.

Spotted grasshopper, *Aularches miliaris*, defends itself with toxic foam and warning colours.

Scouting Techniques

Many agronomists use "foot-square" counts to rate their fields. This is simply the number of grasshopper nymphs that jump from a square foot area as you walk towards it.

To do a field survey, start at one corner of the field, walk diagonally past the centre of the field and then turn and walk straight out to one side of the field. Make at least 20 of these square foot counts in a survey. Dividing the total grasshoppers counted by two will give the approximate number of grasshoppers per square metre (sq. yd.).

Economic Thresholds

When populations exceed eight to 12 hoppers per square metre (sq. yd.), control is usually warranted. At the flowering and podding stage of lentils the threshold is 2/m2.

Control Tips

Early seeded, vigorous crops develop rapidly before the grasshopper eggs hatch. These crops can then tolerate feeding better than late seeded crops.

Hopper baits, containing an insecticide formulated on bran, are another option. Benefits include low cost, no harm to pollinators and most beneficial insects, and avoidance of insecticide drift. The bait can be spread by application equipment designed to spread granular herbicides or seeds, which is available from most agriculture retailers.

Insecticide applications are most effective and the costs are lower when they are applied after the eggs hatch and while the nymphs are still concentrated in their breeding areas.

Often, grasshopper nymphs can be sprayed in roadsides, headlands and field margins before they move into a field crop, vegetable garden or nursery. The lowest dosage given on the container label should be used when the grasshoppers are small and the vegetative cover is low.

Grasshoppers on rangelands may be effectively managed by applying certain insecticides in 30-metre swaths, alternating with 30-metre untreated swaths.

Less preferred crops, such as peas or oats, can be used as guard strips around more preferred crops.

Locust

Locusts are large herbivorous insects that can be serious pests of agriculture due to their ability to form dense and highly mobile swarms. They are species of short-horned grasshoppers that periodically form large populations in dense migrating groups, where individuals differ in several characteristics from those living separately.

Classification of Locusts

Locusts belong to the same order as grasshoppers, katydids and crickets - the Orthoptera (derived from the Greek words *orthos* meaning straight or rigid and *ptera* meaning wing).

In Australia there are over 2,800 species of Orthoptera and over 700 species of locusts and grasshoppers. Many species are yet to be scientifically described.

The following table shows how three main pest species of locusts in Australia are currently classified.

Taxonomic category	Australian plague locust	Migratory locust	Spur-throated locust
Kingdom	Animalia	Animalia	Animalia
Phylum	Arthropoda	Arthropoda	Arthropoda
Superclass	Hexapoda	Hexapoda	Hexapoda
Class	Insecta	Insecta	Insecta

Taxonomic category	Australian plague locust	Migratory locust	Spur-throated locust
Order	Orthoptera	Orthoptera	Orthoptera
Suborder	Caelifera	Caelifera	Caelifera
Superfamily	Acridoidea	Acridoidea	Acridoidea
Family	Acrididae	Acrididae	Acrididae
Subfamily	Oedipodinae	Oedipodinae	Catantopinae
Tribe	-	Locustini	Cyrtacanthacridini*
Genus	Chortoicetes	Locusta	Austracris**
Species	terminifera	migratoria	guttulosa
Subspecies	-	migratorioides	-

* the Cyrtacanthacridini is consider by some taxonomists to be a subfamily of the Acrididae
** *Austracris guttulosa* was formerly classified as *Nomadacris guttulosa*

Cicada

Cicada, (family Cicadidae), any of a group of sound-producing insects (order Homoptera) that have two pairs of membranous wings, prominent compound eyes, and three simple eyes (ocelli). Cicadas are medium to large in size, ranging from 2 to 5 cm (0.8 to 2 inches). Male cicadas produce loud noises by vibrating membranes (timbals) near the base of the abdomen. Most North American cicadas produce rhythmical ticks, buzzes, or whines, although in some species the "song" is musical. Eggs are usually laid in woody plant tissues that drop from the plant when, or shortly after, the eggs hatch. Newly hatched nymphs burrow into the ground where they suck juices from roots of perennialplants. Nymphs usually undergo five molts during the several years required to reach maturity. Although not ordinarily considered a pest, the females, if numerous, may damage young saplings during their egg laying.

Cicada molting

About 1,500 species of cicadas are known. With the exception of two species of hairy cicadas in the family Tettigaretidae that are found only in southeastern Australia, including Tasmania, cicadas belong to the family Cicadidae and are tropical and occur in deserts, grasslands, and forests. In addition to the dog-day cicada (*Tibicen* and others) that appears yearly in midsummer, there are also periodic cicadas. Among the most fascinating and best-known are the 17-year cicada (often erroneously called the 17-year locust) and the 13-year cicada (*Magicicada*). These species occur in large numbers in chronologically and geographically isolated broods.

Magicicada Magicicada, the genus containing the 13- and 17-year cicadas of eastern North America.

The several species are easily recognized by differences in songs, behaviour, and morphology. Males of each species have three distinct sound responses: a congregational song that is regulated by daily weather fluctuations and by songs produced by other males; a courtship song, usually produced prior to copulation; and a disturbance squawk produced by individuals captured, held, or disturbed into flight.

Cicadas have been used in folk medicines, as religious and monetary symbols, and as an important source of food. Their song once was considered to forecast weather changes. In China, male cicadas were caged for their song. The cicada appears in the mythology, literature, and music of many cultures, including that of American Indians.

Cicada Damage to Plants and Trees

These periodical cicadas may attack virtually any tree. The only ones that are fairly safe from their numbers are those that excrete sap, such as pine trees.

Some trees and plants that are most likely to incur damage are apple trees, young grapevines, small trees, and shrubs. Because of this, it is recommended that those who own or plant orchards or nursery know the life cycles of periodical cicadas in there are and not plant young trees or shrubs in a year or two before expected emergence.

Although damage to mature trees may appear extensive after having hundreds or thousands of cicadas living on them, there is generally little lasting damage.

The nymphs that live beneath the ground during the majority of the cicada's life cycle can also cause damage, however, as they feed on the roots of the trees, potentially stunting their growth.

Cicada Harm to Humans

Cicadas do not cause any harm or hazard to humans, as they do not bite, sting, or attack people. If a cicada is handled for too long, it may attempt to feed on a person's hand, but they are not poisonous or known to transmit disease, so the attempt will generally be little more than the feeling of a pinprick.

Natural Cicada Control

- Many animals feed on these pests, however, because they emerge in such high numbers, it is virtually impossible for them to all be attacked or consumed.
- Cicada predators include birds, such as grackles and crows, and fish, which will gorge on the insects. In addition, parasitic wasps, flies, predatory mites feed on the eggs, and birds and cicada killer wasps feed on adults.
- Even humans have been known to dine on cicadas. According to Dr. Gene Kritsky, Professor of Biology at the College of Mount St. Joseph, "Periodical cicadas are best eaten when they are still white, and they taste like cold canned asparagus. Like all insects, cicadas have a good balance of vitamins, are low in fat, and, especially the females, are high in protein." Early Native Americans and John the Baptist were said to eat cicadas.

Non-Chemical Cicada Control

- Young vines/vineyards. cover the trunks of the young vines with aluminum foil or other covering to keep nymphs from climbing up.
- Small or young trees. Wrap the branches loosely with cheesecloth or other mesh cloth. This will keep females from being able to drill into the branches to lay their eggs. Apply the cloth or screening as soon as cicadas emerge and leave it on for about a month, or until the cicadas are gone.
- Do not plant trees within the year or two prior to the periodical cicadas' regional year of emergence.

Chemical Control of Cicadas

- Because the periodical cicadas are so numerous, pesticide application is rarely practical.
- However, there are some insecticides registered for use on deciduous shrubs, fruit, nut, and shade trees that can provide some protection. The first application should be made prior to egg laying; this is approximately 7-10 days after the male periodical cicadas start their singing.

- It may be necessary to repeat the application to keep the female cicadas from laying her eggs.
- When using any pesticide, read and follow all label directions.

Root-knot Nematode

Galls (knots) on roots of a dogwood seedling (*Cornus florida*) infested with root knot nematodes.

Root-knot nematodes are microscopic, plant-parasitic roundworms in the genus, *Meloidogyne*. Normally they exist in sandy soil in hot climates or short winters, but they have spread to our region as well. In Missouri alone 40 species of parasitic nematodes exist. About 2000 plants are susceptible to infection by these nematodes, and they can cause global crop loss. Root-knot nematode larvae infect plant roots that drain the plant's ability to take up water and sufficient nutrients. Because of their microscopic size the nematodes go unnoticed until serious symptoms appear. The pathogenic nematodes can allow entrances for root rots, and wilts. They can also transmit viral diseases.

Symptoms and Diagnosis

A plant infected with root-knot nematodes often exhibits the same signs of nutrient deficiency. It may be stunted, yellowed, or wilted, and can lead to premature death of the plant. Below the ground, the roots of the plant are swollen or knotted with root galls, and unless a plant is removed from the soil to examine the roots, the cause of the poor plant growth remains undetected. The appearance of the galls on the roots can be confused with nitrogen-fixing nodules common in the legume family. However it is easy to distinguish between them, as the nitrogen nodules can be easily rubbed off while the nematode galls cannot be removed.

Life Cycle

The root-knot nematode is parthenogenic, that is a single female can reproduce without males and a new generation can occur every 28 days if conditions are ideal. Inside the gall, the enlarged female appears as a shiny white body, the size of a pinhead. She deposits 300 to 500 eggs in a protective jelly-like material. These glistening white to yellow egg masses are present on the root surfaces. Juveniles emerge from the eggs in the soil and penetrate between and through cells at the center of the root, usually near the growing tip. These larvae actively feed and remain at this same site. The juvenile stage can over-winter even under very unfavorable conditions. New adult nematodes develop from the larvae and start the cycle again.

Damage

Root knot nematodes usually cause distinctive swellings, called galls, on the roots of affected plants. Infestations of these nematodes are fairly easy to recognize; dig up a few plants with symptoms, wash or gently tap the soil from the roots, and examine the roots for galls. The nematodes feed and develop within the galls, which can grow as large as 1 inch in diameter on some plants but usually are much smaller.

The formation of these galls damages the water- and nutrient-conducting abilities of the roots. Galls can crack or split open, especially on the roots of vegetable plants, allowing the entry of soil-borne, disease-causing microorganisms. Root knot nematode galls are true swellings and can't be rubbed off the roots as can the beneficial, nitrogen-fixing nodules on the roots of legumes. Root knot nematodes can feed on the roots of grasses and certain legumes without causing galling.

Aboveground symptoms of a root knot nematode infestation include wilting during the hottest part of the day even with adequate soil moisture, loss of vigor, yellowing leaves, and other symptoms similar to a lack of water or nutrients. Infested vegetable plants grow more slowly than neighboring, healthy plants, beginning in early to midseason. Plants produce fewer and smaller leaves and fruits, and ones heavily infested early in the season can die. Damage is most serious in warm, irrigated, sandy soils.

Root injury from other nematode species can produce aboveground symptoms similar to those from root knot nematodes. However, the actual injury to the roots is more difficult to detect. Roots can be shortened or deformed with no other clues as to the source of the injury. You can confirm a nematode infestation by collecting soil and root samples and sending the material to a laboratory for positive identification of the infesting species.

Although nematodes can kill annual plants, they rarely kill woody plants. Nematode injury to woody plants usually is less obvious and often more difficult to diagnose. Infested fruit and nut trees can have reduced growth and yields. Woody landscape plants that are heavily infested can have reduced growth and branch tip dieback and can defoliate earlier than normal.

Detecting Nematodes in Soil Samples

Table: Landscape Plants Known or Suspected of Being Damaged by Nematodes in California.

Host plant	Nematode(s)[1]
albezia	root knot
alder	root knot
azalea	stunt
boxwood	root knot
cactus	root knot, cyst
catalpa	root knot
cedar	root knot, pinewood
euonymus	root knot
fir	dagger
ginkgo	root knot
hibiscus	root knot
hydrangea	root knot
juniper	root knot
larch	pinewood
lilac	citrus
mulberry	root knot
oak	root knot
palm	root knot
pine	pinewood
pittosporum	root knot
poinsettia	root knot
rose	root knot, root lesion
spruce	pinewood
tamarisk	root knot

[1]Most varieties susceptible to at least one species of the nematode type listed.

Nematodes are too small to see without a microscope. Often you become aware of a nematode problem by finding galled roots on a previous crop. However, you also can use a simple bioassay to detect root knot nematodes in garden soil. Melons seeded in pots in moist soil collected from the garden will develop visible galls on the roots in about 3 weeks when pots are kept at about 80°F if root knot nematodes are present. As a comparison, melons planted in heat-sterilized soil won't develop galls.

Management

Management of nematodes is difficult. The most reliable practices are preventive, including sanitation and choice of plant varieties. You can reduce existing infestations through fallowing, crop rotation, and soil solarization. However, these methods reduce nematodes primarily in the top foot or so of the soil, so they are effective only for about a year. They are suitable primarily for annual plants or to help young woody plants establish. Once nematodes infest an area or crop, try to minimize damage by adjusting planting dates to cooler times of the season when nematodes are less active. Try to provide optimal conditions for plant growth including sufficient irrigation and soil amendments to make plants more tolerant to nematode infestation.

Sanitation

Nematodes usually are introduced into new areas with infested soil or plants. Prevent nematodes from entering your garden by using only nematode-free plants purchased from reliable nurseries. To prevent the spread of nematodes, avoid moving plants and soil from infested parts of the garden. Don't allow irrigation water from around infested plants to run off, as this also spreads nematodes. Nematodes can be present in soil attached to tools and equipment used elsewhere, so clean tools thoroughly before using them in your garden.

Resistant or Tolerant Varieties and Rootstocks

One of the best ways to manage nematodes is to use vegetable varieties and fruit tree rootstocks that are resistant to nematode injury. Tomato varieties with the code VFN (Verticillium, Fusarium, Nematodes) on the seed packet or label are resistant to common root knot nematode species. Although even resistant tomato varieties can still exhibit some root galling under high nematode levels, they usually maintain their yield. For example in recent vegetable garden-type experiments on root knot nematode soil, nematode-resistant tomatoes yielded almost 6 times more tomatoes than a similar susceptible variety. An additional benefit of growing a resistant variety is the nematode levels in the soil decline rather than increase, making it more feasible to grow a susceptible crop the following season.

Table: Fruits and Nuts Known or Suspected of Being Damaged by Nematodes in California.

Host plant	Nematode(s)[1]
Grapes and small fruits	
grape	root knot[2], root lesion, ring, citrus, dagger, stubby root

blackberry, raspberry	root lesion, dagger
strawberry	root knot, root lesion, foliar
Fruit and nut trees	
almond	root knot[3], root lesion, ring
apple	root knot, root lesion
apricot	root knot[4], root lesion[4], ring
avocado	root lesion
cherry	root lesion
citrus	root lesion, citrus
olive	root lesion, citrus
peach, nectarine	root knot[3], root lesion, ring
pear	root lesion
plum, prune	root lesion, ring, pin
walnut	root knot, root lesion

[1] Most varieties susceptible to at least one species of the nematode type listed.
[2] Harmony and Freedom grape rootstocks are resistant to root knot nematodes.
[3] Nemaguard and Nemared (peach) rootstocks are resistant to root knot nematodes.
[4] Royal Blenheim rootstock is resistant to root knot and root lesion nematodes.

For fruit trees and vines, Nemaguard rootstock used for stone fruit and almond trees and Harmony and Freedom rootstock used for grapes provide protection against root knot and other nematodes. Citrus trees growing on Troyer and Trifoliate rootstocks are resistant to the citrus nematode. Consider replacing severely infested plants with plant species and varieties that are more tolerant of the nematodes present. Unfortunately, resistant varieties aren't available for many crops and ornamentals.

Fallowing and Rotation

Growing a crop on which the nematode pest can't reproduce is a good way to control some nematodes. For example, the sugarbeet cyst nematode attacks only a limited number of crops including cole crops (broccoli, Brussels sprouts, cabbage, and cauliflower) and related crops and weeds. Growing nonsusceptible crops for 3 to 5 years reduces the sugarbeet cyst nematode population to a level where you can grow susceptible crops again. Unfortunately, rotation isn't as easy for controlling root knot nematodes, because so many vegetable crops and weeds are hosts of the pest.

However, with careful planning, rotation in combination with fallowing and solarization can reduce root knot nematode numbers. Annual crops that are useful in a rotation plan for reducing root knot nematode populations include small grains such as wheat and barley, sudangrass, and resistant tomato and bean varieties.

Fallowing is the practice of leaving the soil bare for a period of time. Fallowing for 1 year will lower root knot nematode populations enough to successfully grow a susceptible annual crop. Two fallow years will lower nematode numbers even further. When fallowing, it is important to keep the soil moist to induce egg hatch and to control weeds on which nematodes can survive. As a result, eggs will hatch, but the nematodes will die if there is nothing to feed on.

You will need to repeat fallowing when you begin to see root injury again, as nematodes can build up to damaging levels even in a single season. A good way to conduct a fallowing program is to split the garden into thirds and fallow one-third every year or two on a rotating basis. If you intend to grow woody plants in a nematode-infested area, consider fallowing the soil for 4 years before planting. Table below gives an example of a rotation/fallowing plan that would be useful for root knot nematode control.

Soil Solarization

Table: Vegetables Known or Suspected of Being Damaged by Nematodes in California.

Host plant	Nematode(s)[1]
Vegetables	
beans	root knot[2], root lesion
beets	root knot, cyst
carrots	root knot
celery	root knot
cole crops	root knot, cyst
corn	root lesion
cucumbers	root knot
eggplant	root knot
garlic	stem and bulb
lettuce	root knot
melons	root knot
onions	stem and bulb
peas	root knot, root lesion, cyst
peppers	root knot
potatoes (Irish)	root knot, root lesion
potatoes (sweet)	root knot
radish	root knot, cyst

spinach	root knot, cyst
squash	root knot
tomatoes	root knot[3]
turnips	root knot, cyst

[1] Most varieties susceptible to at least one species of the nematode type listed.
[2] Some blackeye, lima, and snap bean varieties are resistant to *Meloidogyne incognita*, a species of root knot nematode.
[3] Tomato varieties designated "N" are resistant to most root knot nematode species.

You can use solarization to temporarily reduce nematode populations in the top 12 inches of soil, which allows the production of shallow-rooted annual crops and helps young woody plants become established before nematode populations increase. However, solarization won't provide long-term protection for fruit trees, vines, and woody ornamental plants.

For effective solarization, moisten the soil, then cover it with a clear, plastic tarp. Leave the tarp in place for 4 to 6 weeks during the hottest part of summer. Root knot nematodes, including eggs, die when soil temperature exceeds 125°F for 30 minutes or 130°F for 5 minutes. The effectiveness of solarization is reduced in cool coastal areas, where summer temperatures commonly remain below 80°F.

Planting and Harvesting Dates

Most nematode species are active during warm summer months and can't penetrate roots at soil temperatures below 64°F. Therefore, you can reduce nematode injury to fall-planted crops such as carrots, lettuce, spinach, and peas by waiting until soil temperatures have dropped below 64°F. Plant summer vegetables as early as possible in spring before nematodes become active. Plants with larger root systems, even though nematode-infested, might be able to remain productive longer. It is also helpful to remove annual vegetables, including their roots, as soon as harvest is over, to prevent nematodes from feeding and breeding on root systems.

Nematode-suppressive Plants

Certain marigolds, *Tagetes* species, suppress root knot and lesion nematodes. French marigolds (varieties include Nemagold, Petite Blanc, Queen Sophia, and Tangerine) are most effective. Avoid signet marigolds, *T. signata* or *tenuifolia*, because nematodes will feed and reproduce on these. Marigolds don't work well against the northern root knot nematode, *Meloidogyne hapla*, a species common in areas with cool winters. The effect of marigolds is greatest when you grow them as a solid planting for an entire season. When grown along with annual vegetables or beneath trees or vines (intercropping), nematode control usually isn't very good. To prevent marigold seed from getting in the soil, cut or mow the plants before the flowers open. As with other cultural control methods, nematode populations rapidly will increase as soon as you grow susceptible crops again.

Soil Amendments and Irrigation

You can add various organic amendments to the soil to reduce the effect of nematodes on crop plants. The amendments—which include peat, manure, and composts—are useful for increasing the water- and nutrient-holding capacity of the soil, especially sandy soils. Because nematodes more readily damage plants that are water-stressed, increasing the soil's capacity to hold water can lessen the effects of nematode injury. Likewise, more frequent irrigation can help reduce nematode damage. In either case, you will have just as many nematodes in the soil, but they will cause less damage.

Pesticides

Currently no chemical nematicides or soil fumigants are available to home gardeners for nematode control in backyard gardens and home landscapes.

Table: Example of a Rotation Plan for a Root Knot Nematode-infested Garden[1].

	First winter	First summer	Second winter	Second summer	Third winter	Third summer
Section A	fallow	fallow	fallow	plant summer-susceptible crop	plant winter/spring crop	plant summer-resistant crop
Section B	plant winter/ spring crop	summer solarize	plant winter/ spring crop	plant summer-susceptible crop	fallow	summer solarize
Section C	plant winter/ spring crop	amend soil, plant summer-susceptible crop	fallow	fallow	fallow	fallow or plant summer-resistant crop

[1]Garden is divided into three sections: A, B, and C.

Soybean Cyst Nematode

The soybean cyst nematode (SCN), *Heterodera glycines,* is a small plant-parasitic roundworm that attacks the roots of soybeans. Most nematodes can be observed only with magnification, but the adult females and cysts of SCN are about '/32 of an inch long and visible to the unaided eye.

Several kinds of plant-parasitic nematodes have been present in Iowa soils since long before the land was used for agriculture. However, SCN is relatively new to Iowa. It probably was introduced into the state 20 to 25 years ago and was first confirmed in Winnebago County in 1978.

The presence of SCN is not usually obvious at the time of initial soil infestation. The nematode population density must increase in the soil until it is sufficient to cause above-ground symptoms on plants or a decrease in yield. Population densities may take several years to reach significant numbers. Thus, current SCN damage to soybeans is the result of infestations that have been growing for several years.

Most likely, SCN came from Japan where it was first reported more than 75 years ago. The first reports of SCN in the United States came from North Carolina in 1954. Since then it has been found in 25 more states in the Southeast and Midwest, including Illinois, Minnesota, Missouri, Nebraska, and Wisconsin.

Symptoms

Above-ground Symptoms

Above-ground symptoms of SCN damage are not unique to SCN. They often are mistaken for damage from compaction, iron deficiency chlorosis and other nutrient deficiencies, drought stress, herbicide injury, or other plant diseases. SCN injury often has remained undetected for several years because these nondescript symptoms were attributed to other causes.

Area of soybean cyst nematode damage in a soybean field.

Rows of stunted, chlorotic soybean plants damaged by soybean cyst nematode are slow to fill in with foliage.

The first obvious symptom of SCN injury to soybeans is the appearance in the field of circular- or oval-shaped areas of stunted, yellowed, less vigorous plants (figure above). These infested areas will vary in their size, often showing a sharp dividing line at the edges between stunted and apparently healthy plants. Plants growing in heavily infested soils may remain stunted throughout the season. Rows of soybeans grown on SCN-infested land frequently are slow to close or fill in with foliage (figure above).

In areas with high pH soils, the yellowing of soybeans due to SCN often is confused with iron deficiency chlorosis. However, there are differences between symptoms of the two problems. Iron deficiency chlorosis symptoms usually appear in early June, whereas yellowing due to SCN will more likely occur in July and August. The yellowing caused by iron deficiency chlorosis typically affects the areas between the veins of the upper leaves. Yellowing due to SCN usually starts at the edges of the leaves, and can affect leaves on the entire plant. Iron deficiency chlorosis and SCN may occur in the same field and even on the same plant.

Adult female of soybean cyst nematode with egg filled egg sac (magnified approximately 25 times).

An area of SCN damage often will appear elongated in the direction of tillage operations. Most severe damage is often in the center of the area, with damage decreasing towards the margins. Such areas frequently develop near a field gate, entrance, wherever equipment enters a field, or near fences where wind-blown soil may accumulate.

The above-ground symptoms of SCN damage do not always occur consistently. Symptoms range from nonexistent to severe depending on the age and vigor of the soybean plants, SCN numbers, soil fertility, moisture levels, and other environmental conditions. Injury usually is more severe in light, sandy soils, but it also occurs in heavier soils. SCN damage is not always confined to smaller areas within a field. Some fields in Iowa are infested with SCN throughout; in such fields, areas of stunted plants are not obvious because the size of plants throughout the field is much the same.

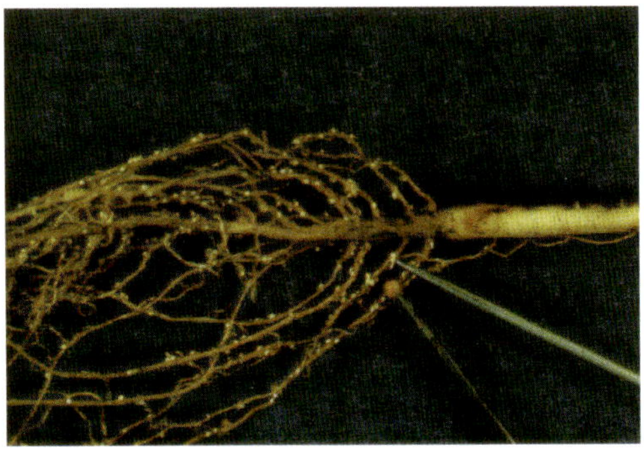

Yellow soybean cyst nematode cysts on soybean roots. Note the small size of the cysts (upper pointer) relative to the larger nitrogen nodule (lower pointer).

One cannot rely upon above-ground symptoms for identification of SCN infestations. If soybean yields in any field have decreased for no apparent reason, or if SCN has been confirmed on nearby land, more thorough examination of plants for below-ground symptoms and a soil analysis for SCN are needed.

Below-ground Symptoms

Most below-ground symptoms of SCN damage are not unique. Roots infected with SCN are dwarfed or stunted. SCN can decrease the number of nitrogen-fixing nodules on the roots. SCN infections also may make the roots more susceptible to attacks by other soil-borne plant pathogens. Often it is difficult to recognize if roots are stunted and have fewer nodules unless they are compared to uninfected soybean plants.

The only unique symptom of SCN infection is the presence of adult female nematodes and cysts on the soybean roots (figure above). These structures, which appear as tiny, lemon-shaped objects on the roots, are white initially, but turn yellow and then tan to brown as they mature. They can be seen with the unaided eye, although observation with a magnifying glass is easier. The cysts are about the size of a pinhead and are much smaller than nitrogen nodules (figure above). Roots must be carefully removed from the soil for examination or the cysts may be dislodged. Observation of adult females and cysts on the roots of soybean plants is the only accurate way to diagnose SCN infestation in the field.

Life Cycle

The SCN life cycle has three major stages: egg, juvenile, and adult. The life cycle can be completed in 24 to 30 days under optimum conditions. When temperature and moisture levels become adequate in the spring, worm-shaped juveniles (figure below) hatch from eggs in the soil. These juveniles are the only life stage of the nematode that can infect soybean roots.

After penetrating the soybean roots, juveniles move through the root until they contact vascular tissue. There they stop and start to feed. The nematode injects secretions that modify certain root cells and transform them into specialized feeding sites. As the nematodes feed, they swell (figure below) and eventually female nematodes becomes so large that they break through the root tissue and are exposed on the surface of the root (figure below).

Male nematodes, which are not swollen as adults, migrate out of the root into the soil and fertilize the lemon-shaped adult females. The males then die, while the females remain attached to the root system and continue to feed. The swollen females begin producing eggs, initially in a mass or egg sac outside the body, then later within the body cavity. Eventually the entire body cavity of the adult female is filled with eggs, and the female nematode dies. It is the egg-filled body of the dead female that is referred to as the cyst. Cysts eventually dislodge and are free in the soil. The walls of the cyst become very tough, providing excellent protection for the 200 to 400 eggs contained within. SCN eggs survive within the cyst until proper hatching conditions occur. Although many of the eggs may hatch within the first year, many also will survive within the cysts for several years.

How SCN Spreads

SCN can move through the soil only a few inches per year on its own power. However, SCN can be spread substantial distances in a variety of ways. Anything that can move infested soil (even small amounts) is capable of spreading SCN, including farm machinery, vehicles and tools, wind, water, animals, and farm workers. Seed sized particles of soil, called soil peds, often contaminate harvested seed. Consequently, SCN can be spread when seed from infested fields is planted in uninfested fields. There is even evidence that SCN can be spread by birds. Only some of these causes can be prevented.

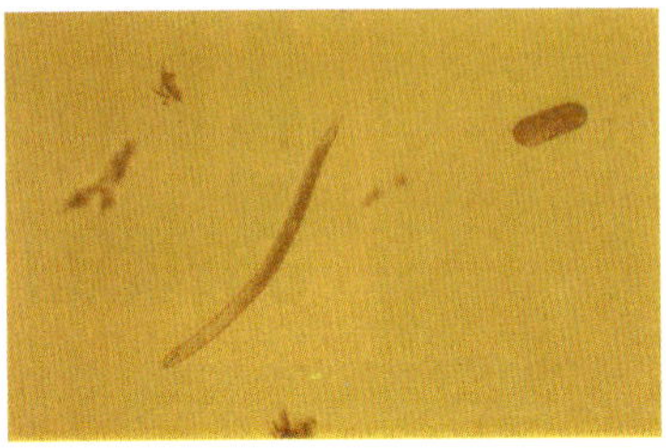

A vermiform, infective soybean cyst nematode juvenile (shortly after hatch) and an unhatched egg (magnified approximately 75 times).

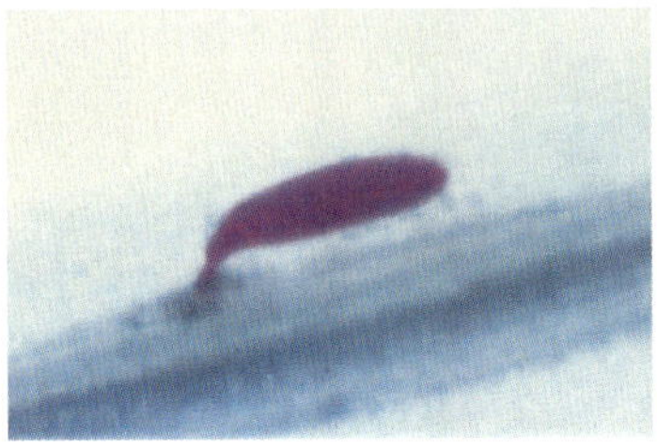

Swollen soybean cyst nematode juvenile within a stained soybean root segment.

Adult soybean cyst nematode females on an infected soybean root segment.

Genetic Resistance

Resistance to SCN has been identified and is available in many soybean varieties. A rotation of resistance sources is recommended for SCN infested fields. Anytime SCN is confirmed in a field, the following rotation should be used: non-host crop – resistant soybean – non-host crop – resistant soybean. Additional years of the non-host crop will reduce the SCN population (number of eggs) further. When different SCN resistance sources cannot be identified (PI88788 is the most common), a change in resistant soybean variety should be substituted. All varieties with PI88788 are not the same as to their effect on SCN.

Cultural Practices

Yield loss due to SCN will be reduced by maintaining optimum growing conditions and avoiding plant stress. Moisture stress is the most common factor increasing SCN losses. Avoid spreading SCN by working and planting infested fields last. When possible, clean equipment before moving to the next field to minimize movement potential. Anything that moves soil will spread SCN.

Potato Cyst Nematode

Potato cyst nematode (PCN) is a serious pest of potatoes world-wide and is subject to stringent quarantine and/or regulatory procedures wherever it occurs. PCN can be a devastating pest of potatoes in temperate regions if not controlled. It was introduced into Europe from South America in the mid-1880s and was first discovered in Australia in 1986 at Munster in Western Australia. PCN was first found in Victoria in 1991 on potato crops in market gardens in the Wandin-Silvan area and subsequently in several other areas. There are two species of PCN, *Globodera rostochiensis* and *Globodera pallida*, but only *G. rostochiensis* is known to occur in Australia. Only one of the strains of *G. rostochiensis*, the Ro1 strain, has ever been confirmed in Australia.

Symptoms

Infested plants are stunted and may wilt (figure below), leaves may yellow or display a dull colour. Affected plants have a reduced root system which is abnormally branched and brownish in colour. At flowering or later, minute-white, yellow or brown spheres or cysts, about the size of a pin head (0.5 mm), can be seen on the outside of roots.

Damage to the crop varies from small patches of poor growing plants to complete crop failure. Diseased plants first occur in isolated patches and these become larger with each new crop if potatoes are continually grown on the infested site. In light infestations, potato plants may show no above ground symptoms, but yield can be reduced. Light infestations can reduce tuber size, whereas heavy infestations reduce both number and size of tubers. PCN is not greatly influenced by soil type and temperature because the nematode thrives wherever potatoes are grown.

Symptoms of PCN infestation in a Victorian market potato crop. Plants on the left-hand side are stunted due to a heavy infestation of PCN.

PCN infested potato roots under high magnification. The cysts are initially yellow and darken and tan as they age. Cysts are the bodies of female nematodes and contain the nematode eggs.

Description

Nematodes or eelworms are small worm-shaped organisms, less than 1 mm in length, which inhabit soil and attack plant roots. The cysts (figure above) of *G. rostochiensis* (golden nematode) are white, yellow or golden in colour when they first form on roots (immature cysts) and become tan brown in colour when they mature.

Control

Because PCN cysts are extremely long lived in soil, even in the absence of suitable host plants, eradication of PCN from an infested site is not possible in the short term.

However, eradication remains as the long term objective for all infested land. Overseas studies have shown that cysts may survive longer than 20 years, even in the absence of suitable host plants. Ongoing management of PCN outbreak sites should aim to progressively reduce PCN populations, eventually to the point where PCN may be considered to be eradicated. Overseas there is a major emphasis on the use of resistant cultivars to suppress soil populations of PCN. In Victoria, and Australia, there are a range of regulatory mechanisms which aim to minimise the spread of PCN and reduce populations on infested land. To reduce PCN populations and minimise the risk of spread of PCN, both within and outside a property, a farm management plan should consider appropriate biosecurity strategies. These include:

- use of certified seed
- soil testing for PCN
- hygiene protocols for potatoes, equipment, farm workers, soil and water
- resistant potato cultivars
- control of non resistant self sown potatoes
- crop rotations
- trap cropping (not yet done in Australia as *G. pallida* is not yet known to occur)
- nematicides and soil solarisation
- alternative crops.

Termites

Termites usually are cryptic, meaning that they don't come out into the open making them difficult to detect. They are often called the "silent destroyer" because they may be secretly hiding and thriving in your home or yard without any immediate signs of damage. Even when they are visible within the home as flying insects, termites can often be mistaken for ants. Some homeowners will dismiss the termites as pesky ants (which often swarm at the same time of year) and ignore them without taking any preventive or extermination measures. This allows established termite infestations to grow exponentially. Termites can be identified by their straight, beaded antennae, uniform waists, and wings of equal size. (Ants have elbowed antennae, constricted waists and forewings that are longer than the hind wings.)

While termite workers only measure approximately 1 cm to a few millimeters in length, their feeding habits are capable of causing costly damage to property.

Termites primarily feed on cellulose, a part of wood, but also damage paper, books, insulation, and even swimming pool liners and filtration systems. All termites feed on detritus (dead plants and trees as well as dead parts of living trees, including wood). Unfortunately, all homes, regardless of their construction type, can provide cellulose food for termite infestation. House foundations, furniture, shelves and even books are all possible feeding sites for termites. Termites can harm living trees and shrubs, but are more often are a secondary invader of woody plants already in decline.

Termite Characteristics

Termites live in large social groups called colonies. These colonies are hungry, too, eating 24 hours a day, every day. Termites aren't just abundant; they are also varied as there are about 2,000 known species on Earth. In the United States alone there are over 40 different species.

Termites are insects, so they have body segments and six legs. They may also have wings, and the wings are all about the same length. Termites are often confused with ants, but you can tell the difference because an ant's front wings are longer than its hind wings. Also, a termite's antennae will come straight out of its head, while an ant's will bend at a right angle. Termites aren't very big, just ¼ - ½ inch in length. Termite kings and queens of some species may be longer though, up to a full inch. Termites are small, but because they are so plentiful, the total weight of all the termites is more than the total weight of all the humans on Earth.

Formosan termites

A termite swarm outside is no problem, but inside your house is a big deal!

The swarmers of the colony are the winged termites, and they are tasked with reproducing and creating new termite colonies. Workertermites, on the other hand, are tasked with, you guessed it, working. They are sterile, so they can't reproduce. But they make up the majority of the colony and are critical because they feed the colony. These are the ones that are in your walls feeding away. Swarmers outside your house are not a problem, but if you see swarmers in your house (or their wings, since they fall off quickly once inside), then you're dealing with an established colony.

Behaviour and Ecology

Diet

Termite faecal pellets

Termites are detritivores, consuming dead plants at any level of decomposition. They also play a vital role in the ecosystem by recycling waste material such as dead wood, faeces and plants. Many species eat cellulose, having a specialised midgut that breaks down the fibre. Termites are considered to be a major source (11%) of atmospheric methane, one of the prime greenhouse gases, produced from the breakdown of cellulose. Termites rely primarily upon symbiotic protozoa (metamonads) and other microbes such as flagellate protists in their guts to digest the cellulose for them, allowing them to absorb the end products for their own use. Gut protozoa, such as *Trichonympha*, in turn, rely on symbiotic bacteria embedded on their surfaces to produce some of the necessary digestive enzymes. Most higher termites, especially in the family Termitidae, can produce their own cellulase enzymes, but they rely primarily upon the bacteria. The flagellates have been lost in Termitidae. Scientists' understanding of the relationship between the termite digestive tract and the microbial endosymbionts is still rudimentary; what is true in all termite species, however, is that the workers feed the other members of the colony with substances derived from the digestion of plant material, either from the mouth or anus. Judging from closely related bacterial species, it is strongly presumed that the termites' and cockroach's gut microbiota derives from their dictyopteran ancestors.

Certain species such as *Gnathamitermes tubiformans* have seasonal food habits. For example, they may preferentially consume Red three-awn (*Aristida longiseta*) during the summer, Buffalograss (*Buchloe dactyloides*) from May to August, and blue grama *Bouteloua gracilis* during spring, summer and autumn. Colonies of *G. tubiformans* consume less food in spring than they do during autumn when their feeding activity is high.

Various woods differ in their susceptibility to termite attack; the differences are attributed to such factors as moisture content, hardness, and resin and lignin content. In one study, the drywood termite *Cryptotermes brevis* strongly preferred poplar and maple woods to other woods that were generally rejected by the termite colony. These preferences may in part have represented conditioned or learned behaviour.

Some species of termite practice fungiculture. They maintain a "garden" of specialised fungi of genus *Termitomyces*, which are nourished by the excrement of the insects. When the fungi are eaten, their spores pass undamaged through the intestines of the termites to complete the cycle by germinating in the fresh faecal pellets. Molecular evidence suggests that the family Macrotermitinae developed agriculture about 31 million years ago. It is assumed that more than 90 percent of dry wood in the semiarid savannah ecosystems of Africa and Asia are reprocessed by these termites. Originally living in the rainforest, fungus farming allowed them to colonise the African savannah and other new environments, eventually expanding into Asia.

Depending on their feeding habits, termites are placed into two groups: the lower termites and higher termites. The lower termites predominately feed on wood. As wood is difficult to digest, termites prefer to consume fungus-infected wood because it is easier to digest and the fungi are high in protein. Meanwhile, the higher termites consume a wide variety of materials, including faeces, humus, grass, leaves and roots. The gut in the lower termites contains many species of bacteria along with protozoa, while the higher termites only have a few species of bacteria with no protozoa.

Predators

Termites are consumed by a wide variety of predators. One termite species alone, *Hodotermes mossambicus*, was found in the stomach contents of 65 birds and 19 mammals. Ants, arthropods, reptiles, and amphibians such as bees, centipedes, cockroaches, crickets, dragonflies, frogs, lizards, scorpions, spiders, and toads consume these insects, while 2 spiders in the family Ammoxenidae are specialist termite predators. Other predators include aardvarks, aardwolves, anteaters, bats, bears, bilbies, many birds, echidnas, foxes, galagos, numbats, mice and pangolins. The aardwolf is an insectivorous mammal that primarily feeds on termites; it locates its food by sound and also by detecting the scent secreted by the soldiers; a single aardwolf is capable of consuming thousands of termites in a single night by using its long, sticky tongue. Sloth bears break open mounds to consume the nestmates, while chimpanzees have developed tools to "fish" termites from their nest. Wear pattern analysis of bone tools used by the early hominin *Paranthropus robustus* suggests that they used these tools to dig into termite mounds.

Crab spider with a captured alate

A Matabele ant (*Megaponera analis*) kills a *Macrotermes bellicosus*termite soldier during a raid.

Among all predators, ants are the greatest enemy to termites. Some ant genera are specialist predators of termites. For example, *Megaponera* is a strictly termite-eating (termitophagous) genus that perform raiding activities, some lasting several hours. *Paltothyreus tarsatus* is another termite-raiding species, with each individual stacking as many termites as possible in its mandibles before returning home, all the while recruiting additional nestmates to the raiding site through chemical trails. The Malaysian basicerotine ants *Eurhopalothrix heliscata* uses a different strategy of termite hunting by pressing themselves into tight spaces, as they hunt through rotting wood housing termite colonies. Once inside, the ants seize their prey by using their short but sharp mandibles. *Tetramorium uelense* is a specialised predator species that feeds on small termites. A scout recruits 10–30 workers to an area where termites are present, killing them by immobilising them with their stinger. *Centromyrmex* and *Iridomyrmex* colonies sometimes nest in termite mounds, and so the termites are preyed on by these ants. No evidence for any kind of relationship (other than a predatory one) is known. Other ants, including *Acanthostichus, Camponotus, Crematogaster, Cylindromyrmex, Leptogenys, Odontomachus, Ophthalmopone, Pachycondyla, Rhytidoponera, Solenopsis* and *Wasmannia*, also prey on termites. In contrast to all these ant species, and despite their enormous diversity of prey, *Dorylus ants* rarely consume termites.

Ants are not the only invertebrates that perform raids. Many sphecoid wasps and several species including *Polybia Lepeletier* and *Angiopolybia Araujo* are known to raid termite mounds during the termites' nuptial flight.

Parasites, Pathogens and Viruses

Termites are less likely to be attacked by parasites than bees, wasps and ants, as they are usually well protected in their mounds. Nevertheless, termites are infected by a variety of parasites. Some of these include dipteran flies, *Pyemotes* mites, and a large number of nematode parasites. Most nematode parasites are in the order Rhabditida; others are in the genus *Mermis, Diplogaster aerivora* and *Harteria gallinarum*. Under imminent threat of an attack by parasites, a colony may migrate to a new location. Fungi pathogens such as *Aspergillus nomius* and *Metarhizium anisopliae* are, however, major threats to a termite colony as they are not host-specific and may infect large portions of the colony; transmission usually occurs via direct physical contact. *M. anispliae* is known to weaken the termite immune system. Infection with *A. nomius* only occurs when a colony is under great stress.

Termites are infected by viruses including Entomopoxvirinae and the Nuclear Polyhedrosis Virus.

Locomotion and Foraging

Because the worker and soldier castes lack wings and thus never fly, and the reproductives use their wings for just a brief amount of time, termites predominantly rely upon their legs to move about.

Foraging behaviour depends on the type of termite. For example, certain species feed on the wood structures they inhabit, and others harvest food that is near the nest. Most workers are rarely found out in the open, and do not forage unprotected; they rely on sheeting and runways to protect them from predators. Subterranean termites construct tunnels and galleries to look for food, and workers who manage to find food sources recruit additional nestmates by depositing a phagostimulant pheromone that attracts workers. Foraging workers use semiochemicals to communicate with each other, and workers who begin to forage outside of their nest release trail pheromones from their sternal glands. In one species, *Nasutitermes costalis*, there are three phases in a foraging expedition: first, soldiers scout an area. When they find a food source, they communicate to other soldiers and a small force of workers starts to emerge. In the second phase, workers appear in large numbers at the site. The third phase is marked by a decrease in the number of soldiers present and an increase in the number of workers. Isolated termite workers may engage in Lévy flight behaviour as an optimised strategy for finding their nestmates or foraging for food.

Competition

Competition between two colonies always results in agonistic behaviour towards each other, resulting in fights. These fights can cause mortality on both sides and, in some cases, the gain or loss of territory. "Cemetery pits" may be present, where the bodies of dead termites are buried.

Studies show that when termites encounter each other in foraging areas, some of the termites deliberately block passages to prevent other termites from entering. Dead termites from other colonies found in exploratory tunnels leads to the isolation of the area and thus the need to construct new tunnels. Conflict between two competitors does not always occur. For example, though they might block each other's passages, colonies of *Macrotermes bellicosus* and *Macrotermes subhyalinus* are not always aggressive towards each other. Suicide cramming is known in *Coptotermes formosanus*. Since *C. formosanus* colonies may get into physical conflict, some termites squeeze tightly into foraging tunnels and die, successfully blocking the tunnel and ending all agonistic activities.

Among the reproductive caste, neotenic queens may compete with each other to become the dominant queen when there are no primary reproductives. This struggle among the queens leads to the elimination of all but a single queen, which, with the king, takes over the colony.

Ants and termites may compete with each other for nesting space. In particular, ants that prey on termites usually have a negative impact on arboreal nesting species.

Communication

Most termites are blind, so communication primarily occurs through chemical, mechanical and pheromonal cues. These methods of communication are used in a variety of activities, including foraging, locating reproductives, construction of nests, recognition of nestmates, nuptial flight, locating and fighting enemies, and defending the nests. The most common way of communicating is through antennation. A number of pheromones are known, including contact pheromones (which are transmitted when workers are engaged in trophallaxis or grooming) and alarm, trail and sex pheromones. The alarm pheromone and other defensive chemicals are secreted from the frontal gland. Trail pheromones are secreted from the sternal gland, and sex pheromones derive from two glandular sources: the sternal and tergal glands. When termites go out to look for food, they forage in columns along the ground through vegetation. A trail can be identified by the faecal deposits or runways that are covered by objects. Workers leave pheromones on these trails, which are detected by other nestmates through olfactory receptors. Termites can also communicate through mechanical cues, vibrations, and physical contact. These signals are frequently used for alarm communication or for evaluating a food source.

Hordes of *Nasutitermes* on a march for food, following, and leaving, trail pheromones

When termites construct their nests, they use predominantly indirect communication. No single termite would be in charge of any particular construction project. Individual termites react rather than think, but at a group level, they exhibit a sort of collective cognition. Specific structures or other objects such as pellets of soil or pillars cause termites to start building. The termite adds these objects onto existing structures, and such behaviour encourages building behaviour in other workers. The result is a self-organised process whereby the information that directs termite activity results from changes in the environment rather than from direct contact among individuals.

Termites can distinguish nestmates and non-nestmates through chemical communication and gut symbionts: chemicals consisting of hydrocarbons released from the cuticle allow the recognition of alien termite species. Each colony has its own distinct odour. This odour is a result of genetic and environmental factors such as the termites' diet and the composition of the bacteria within the termites' intestines.

Defence

Termites rely on alarm communication to defend a colony. Alarm pheromones can be released when the nest has been breached or is being attacked by enemies or potential pathogens. Termites always avoid nestmates infected with *Metarhizium anisopliae* spores, through vibrational signals released by infected nestmates. Other methods of defence include intense jerking and secretion of fluids from the frontal gland and defecating faeces containing alarm pheromones.

In some species, some soldiers block tunnels to prevent their enemies from entering the nest, and they may deliberately rupture themselves as an act of defence. In cases where the intrusion is coming from a breach that is larger than the soldier's head, defence requires a special formations where soldiers form a phalanx-like formation around the breach and bite at intruders. If an invasion carried out by *Megaponera analis* is successful, an entire colony may be destroyed, although this scenario is rare.

To termites, any breach of their tunnels or nests is a cause for alarm. When termites detect a potential breach, the soldiers usually bang their heads, apparently to attract other soldiers for defence and to recruit additional workers to repair any breach. Additionally, an alarmed termite bumps into other termites which causes them to be alarmed and to leave pheromone trails to the disturbed area, which is also a way to recruit extra workers.

Termites rush to a damaged area of the nest.

Nasute termite soldiers on rotten wood

The pantropical subfamily Nasutitermitinae has a specialised caste of soldiers, known as nasutes, that have the ability to exude noxious liquids through a horn-like frontal projection that they use for defence. Nasutes have lost their mandibles through the course of evolution and must be fed by workers. A wide variety of monoterpene hydrocarbon solvents have been identified in the liquids that nasutes secrete. Similarly, Formosan subterranean termites have been known to secrete naphthalene to protect their nests.

Soldiers of the species *Globitermes sulphureus* commit suicide by autothysis – rupturing a large gland just beneath the surface of their cuticles. The thick, yellow fluid in the gland becomes very sticky on contact with the air, entangling ants or other insects which are trying to invade the nest. Another termite, *Neocapriterme taracua*, also engages in suicidal defence. Workers physically unable to use their mandibles while in a fight form a pouch full of chemicals, then deliberately rupture themselves, releasing toxic chemicals that paralyse and kill their enemies. The soldiers of the neotropical termite family Serritermitidae have a defence strategy which involves front gland autothysis, with the body rupturing between the head and abdomen. When soldiers guarding nest entrances are attacked by intruders, they engage in autothysis, creating a block that denies entry to any attacker.

Workers use several different strategies to deal with their dead, including burying, cannibalism, and avoiding a corpse altogether. To avoid pathogens, termites occasionally engage in necrophoresis, in which a nestmate carries away a corpse from the colony to dispose of it elsewhere. Which strategy is used depends on the nature of the corpse a worker is dealing with (i.e. the age of the carcass).

Relationship with Other Organisms

A species of fungus is known to mimic termite eggs, successfully avoiding its natural predators. These small brown balls, known as "termite balls", rarely kill the eggs, and in some cases the workers tend to them. This fungus mimics these eggs by producing a cellulose-digesting enzyme known as glucosidases. A unique mimicking behaviour exists between various species of *Trichopsenius* beetles and certain termite species within *Reticulitermes*. The beetles share the same cuticle hydrocarbons as the termites and even biosynthesize them. This chemical mimicry allows the beetles to integrate themselves within the termite colonies. The developed appendages on the physogastric abdomen of *Austrospirachtha mimetes* allows the beetle to mimic a termite worker.

Rhizanthella gardneri is the only orchid known to be pollinated by termites.

Some species of ant are known to capture termites to use as a fresh food source later on, rather than killing them. For example, *Formica nigra* captures termites, and those who try to escape are immediately seized and driven underground. Certain species of ants in the subfamily Ponerinae conduct these raids although other ant species go in alone to steal the eggs or nymphs. Ants such as *Megaponera analis* attack the outside of mounds and Dorylinae ants attack underground. Despite this, some termites and ants can coexist peacefully. Some species of termite, including *Nasutitermes corniger*, form associations with certain ant species to keep away predatory ant species. The earliest known association between *Azteca* ants and *Nasutitermes* termites date back to the Oligocene to Miocene period.

An ant raiding party collecting *Pseudocanthotermes militaris* termites after a successful raid

54 species of ants are known to inhabit *Nasutitermes* mounds, both occupied and abandoned ones. One reason many ants live in *Nasutitermes* mounds is due to the termites frequent occurrence in their geographical range; another is to protect themselves from floods. *Iridomyrmex* also inhabits termite mounds although no evidence for any kind of relationship (other than a predatory one) is known. In rare cases, certain species of termites live inside active ant colonies. Some invertebrate organisms such as beetles, caterpillars, flies and millipedes are termitophiles and dwell inside termite colonies (they are unable to survive independently). As a result, certain beetles and flies have evolved with their hosts. They have developed a gland that secrete a substance that attracts the workers by licking them. Mounds may also provide shelter and warmth to birds, lizards, snakes and scorpions.

Termites are known to carry pollen and regularly visit flowers, so are regarded as potential pollinators for a number of flowering plants. One flower in particular, *Rhizanthella gardneri*, is regularly pollinated by foraging workers, and it is perhaps the only Orchidaceae flower in the world to be pollinated by termites.

Many plants have developed effective defences against termites. However, seedlings are vulnerable to termite attacks and need additional protection, as their defence mechanisms only develop when they have passed the seedling stage. Defence is typically achieved by secreting antifeedant chemicals into the woody cell walls.This reduces the ability of termites to efficiently digest the cellulose. A commercial product, "Blockaid", has been developed in Australia that uses a range of plant extracts to create a paint-on nontoxic termite barrier for buildings. An extract of a species of Australian figwort, *Eremophila*, has been shown to repel termites; tests have shown that termites are strongly repelled by the toxic material to the extent that they will starve rather than consume the food. When kept close to the extract, they become disoriented and eventually die.

Relationship with the Environment

Termite populations can be substantially impacted by environmental changes including those caused by human intervention. A Brazilian study investigated the termite assemblages of three sites of Caatingaunder different levels of anthropogenic disturbance in the semi-arid region of northeastern Brazil were sampled using 65 x 2 m transects. A total of 26 species of termites were present in the three sites, and 196 encounters were recorded in the transects. The termite assemblages were considerably different among sites, with a conspicuous reduction in both diversity and abundance with increased disturbance, related to the reduction of tree density and soil cover, and with the intensity of trampling by cattle and goats. The wood-feeders were the most severely affected feeding group.

Nests

A termite nest can be considered as being composed of two parts, the inanimate and the animate. The animate is all of the termites living inside the colony, and the inanimate part is the structure itself, which is constructed by the termites. Nests can be broadly separated into three main categories: subterranean (completely below ground), epigeal (protruding above the soil surface), and arboreal (built above ground, but always connected to the ground via shelter tubes). Epigeal nests (mounds) protrude from the earth with ground contact and are made out of earth and mud. A nest has many functions such as providing a protected living space and providing shelter against predators. Most termites construct underground colonies rather than multifunctional nests and mounds. Primitive termites of today nest in wooden structures such as logs, stumps and the dead parts of trees, as did termites millions of years ago.

An arboreal termite nest in Mexico

Termite nest in a Banksia, Palm Beach, Sydney.

To build their nests, termites primarily use faeces, which have many desirable properties as a construction material. Other building materials include partly digested plant material, used in carton nests (arboreal nests built from faecal elements and wood), and soil, used in subterranean nest and mound construction. Not all nests are visible, as many nests in tropical forests are located underground. Species in the subfamily Apicotermitinae are good examples of subterranean nest builders, as they only dwell inside tunnels. Other termites live in wood, and tunnels are constructed as they feed on the wood. Nests and mounds protect the termites' soft bodies against desiccation, light, pathogens and parasites, as well as providing a fortification against predators. Nests made out of carton are particularly weak, and so the inhabitants use counter-attack strategies against invading predators.

Arboreal carton nests of mangrove swamp-dwelling *Nasutitermes* are enriched in lignin and depleted in cellulose and xylans. This change is caused by bacterial decay in the gut of the termites: they use their faeces as a carton building material. Arboreal termites nests can account for as much as 2% of above ground carbon storage in Puerto Ricanmangrove swamps. These *Nasutitermes* nests are mainly composed of partially biodegraded wood material from the stems and branches of mangrove trees, namely, *Rhizophora mangle* (red mangrove), *Avicennia germinans* (black mangrove) and *Laguncularia racemose* (white mangrove).

Some species build complex nests called polycalic nests; this habitat is called polycalism. Polycalic species of termites form multiple nests, or calies, connected by subterranean chambers. The termite genera *Apicotermes* and *Trinervitermes* are known to have polycalic species. Polycalic nests appear to be less frequent in mound-building species although polycalic arboreal nests have been observed in a few species of *Nasutitermes*.

Mounds

Nests are considered mounds if they protrude from the earth's surface. A mound provides termites the same protection as a nest but is stronger. Mounds located in areas with torrential and continuous rainfall are at risk of mound erosion due to their clay-rich construction. Those made from carton can provide protection from the rain, and in fact can withstand high precipitation. Certain areas in mounds are used as strong points in case of a breach. For example, *Cubitermes* colonies build narrow tunnels used as strong points, as the diameter of the tunnels is small enough for soldiers to block. A highly protected chamber, known as the "queens cell", houses the queen and king and is used as a last line of defence.

Species in the genus *Macrotermes* arguably build the most complex structures in the insect world, constructing enormous mounds. These mounds are among the largest in the world, reaching a height of 8 to 9 metres (26 to 29 feet), and consist of chimneys, pinnacles and ridges. Another termite species, *Amitermes meridionalis*, can build nests

3 to 4 metres (9 to 13 feet) high and 2.5 metres (8 feet) wide. The tallest mound ever recorded was 12.8 metres (42 ft) long found in the Democratic Republic of the Congo.

Cathedral mounds in the Northern Territory, Australia

Mounds of "compass" or "magnetic" termites (*Amitermes*) oriented north-south, thereby avoiding mid-day heat

Termite mound in Queensland, Australia

Termite mound in Namibia

The sculptured mounds sometimes have elaborate and distinctive forms, such as those of the compass termite (*Amitermes meridionalis* and *A. laurensis*), which builds tall, wedge-shaped mounds with the long axis oriented approximately north–south, which gives them their common name. This orientation has been experimentally shown to assist thermoregulation. The north-south orientation causes the internal temperature of a mound to increase rapidly during the morning while avoiding overheating from the midday sun. The temperature then remains at a plateau for the rest of the day until the evening.

Shelter Tubes

Termites construct shelter tubes, also known as earthen tubes or mud tubes, that start from the ground. These shelter tubes can be found on walls and other structures.Constructed by termites during the night, a time of higher humidity, these tubes provide protection to termites from potential predators, especially ants. Shelter tubes also provide high humidity and darkness and allow workers to collect food sources that cannot be accessed in any other way. These passageways are made from soil and faeces and are normally brown in colour. The size of these shelter tubes depends on the amount of food sources that are available. They range from less than 1 cm to several cm in width, but may extend dozens of metres in length.

Nasutiterminae shelter tubes on a tree trunk provide cover for the trail from nest to forest floor.

Relationship with Humans

As Pests

Owing to their wood-eating habits, many termite species can do great damage to unprotected buildings and other wooden structures. Their habit of remaining concealed often results in their presence being undetected until the timbers are severely damaged, leaving a thin layer of a wall that protects them from the environment. Of the 3,106 species known, only 183 species cause damage; 83 species cause significant damage to wooden structures. In North America, nine subterranean species are pests; in Australia, 16 species have an economic impact; in the Indian subcontinent 26 species are considered pests, and in tropical Africa, 24. In Central America and the West Indies, there are 17 pest species. Among the termite genera, *Coptotermes* has the highest number of pest species of any genus, with 28 species known to cause damage. Less than 10% of drywood termites are pests, but they infect wooden structures and furniture in tropical, subtropical and other regions. Dampwood termites only attack lumber material exposed to rainfall or soil.

Termite damage on external structure

Drywood termites thrive in warm climates, and human activities can enable them to invade homes since they can be transported through contaminated goods, containers and ships. Colonies of termites have been seen thriving in warm buildings located in cold regions. Some termites are considered invasive species. *Cryptotermes brevis*, the most widely introduced invasive termite species in the world, has been introduced to all the islands in the West Indies and to Australia.

Termite damage in wooden house stumps

In addition to causing damage to buildings, termites can also damage food crops. Termites may attack trees whose resistance to damage is low but generally ignore fast-growing plants. Most attacks occur at harvest time; crops and trees are attacked during the dry season.

The damage caused by termites costs the southwestern United States approximately $1.5 billion each year in wood structure damage, but the true cost of damage worldwide cannot be determined. Drywood termites are responsible for a large proportion of the damage caused by termites.

To better control the population of termites, various methods have been developed to track termite movements. One early method involved distributing termite bait laced with immunoglobulin G (IgG) marker proteins from rabbits or chickens. Termites collected from the field could be tested for the rabbit-IgG markers using a rabbit-IgG-specific assay. More recently developed, less expensive alternatives include tracking the termites using egg white, cow milk, or soy milk proteins, which can be sprayed on termites in the field. Termites bearing these proteins can be traced using a protein-specific ELISA test.

Thrips

Thrips, (order Thysanoptera), any of approximately 5,000 species of insects that are among the smallest of the winged insects and are abundant in the tropical and temperate regions of the world. Thrips are economically important since some species transmit plant viruses. Feeding by thrips may reduce seed production, disfigure flowers and fruits, and damage plant leaves. In warm areas a number of them cause plant galls and leaf rolls. On the other hand, a few species prey on destructive mites and scale insects, and a number may aid in the pollination of flowers.

Most thrips are 1.5 to 3 mm (0.06 to 0.12 inch) in length, with the smallest being about 0.6 mm (0.02 inch) and the largest about 15 mm (0.6 inch). Because of their minute size, they can enter the smallest flowers or tiniest cracks in stems and bark. Their wings, when present, are narrow and fringed, hence the name Thysanoptera (Latin for "fringed wings"). Despite their Lilliputian size, thrips are often elaborately ornamented with hexagonal designs, spines, or body flanges. The bristles (setae) of some are expanded at the tips or elongated, or the body wall may be expanded into prongs or forks. The larvae of thrips tend to be brightly coloured (red, orange, or yellow), whereas the adults range from whitish to brown or black. Adults have extensible bladders on the tarsi of the legs, as suggested by the common name bladder feet. The term *thrips* is both singular and plural.

Life Cycle

The egg stage is followed by two larval stages, during which all the immature feeding occurs. The development of thrips is unusual in that, although they belong to a group of insects that usually undergo simple metamorphosis, they have pupae resembling those that occur in insects that undergo complete metamorphosis. There may be one, two, or three pupal stages in which no feeding occurs.

The primary method of reproduction among thrips is by bisexual union. The males frequently are shorter lived than females, and mating often takes place early in the life span, when the adults first appear. Reproduction by unfertilized eggs (parthenogenesis) has been known to occur occasionally.

Thrips in the suborder Terebrantia always lay eggs, as do most thrips in the more advanced suborder Tubulifera. Hatching of eggs within the female (ovoviparity), occurs in some highly evolved members of the Tubulifera. Eggs are elongated, oval to kidney shaped, faintly to distinctly sculptured, and, as is the case in other minute insects, are proportionately large, often occupying two or three abdominal segments of the female before laying. Ordinarily, eggs of the Terebrantia are inserted into living plant tissue in a slit prepared by the sawlike ovipositor of the female. By contrast, eggs of the Tubulifera, whose females have no ovipositor, are pushed under bark, in or between flowers and leaf sheaths, or are glued onto leaf or bark surfaces. Eggs take from 3 to 16 days to hatch according to the species and conditions.

Although similar in overall form to the adult, the larvae of thrips are smaller and simpler in structure. The long anal bristles characteristic of larvae may hold droplets of excreted liquid that are deposited on leaf surfaces. The presence of dark-coloured spots of excreta adjacent to light-coloured feeding zones are some typical signs of thrips damage.

The first larval stage is short, lasting only a few days. The second may last up to many months when estivation or hibernation occurs. In the Terebrantia the second stage larva transforms into a prepupa, whereas in the Tubulifera an abbreviated form, the primipupa, is part of the sequence of development and is followed by the prepupal form. The prepupal stage lasts only a few hours to several days and gives rise to the final pupal stage, which may be of short or long duration. Pupation takes place in the ground in earthen cells (in the case of many advanced Terebrantia), in cocoons (in many primitive Terebrantia), or on a host plant (in many advanced Terebrantia and the Tubulifera). Some thrips, for example, the ear thrips, have only one generation per year, whereas others such as the onion thrips may have several generations each growing season.

Ecology

In regions having cold winters, thrips hibernate as adults or immature forms in detritus, hollow stems, or in the ground. Those that perish with the cold are replaced each year by spring migrants from warmer areas. In the tropics, many thrips are more active in the wet than in the dry seasons. Some thrips are sensitive to relative humidity. Although larvae are generally found in shady locations, on the underside of leaves or branches, adults occasionally occur on the upper side of leaves in bright light. Most thrips rest tightly against leaf veins or in crevices. They are primarily active during the daylight hours.

Thrips are preyed upon by many insects (including other thrips), mites, birds, salamanders, and lizards. However, heavy rains, winds, and dust are probably as destructive to thrips as are predators.

The Larva

In the larvae, the compound eyes are composed of a few facets, simple eyes (ocelli) are absent, antennal segments are fewer than in the adult, the middle portion of the body (thorax) is simple, the lower part of the leg (tarsus) is always one segmented (although also occasionally so in some adults), and the wings and genitalia are not developed. Larval bristles (setae) often differ in form, number, and arrangement from adult setae. Occasionally additional processes such as spines, combs, and plates are found on larvae.

The Adult

The head of a thrips is slightly skewed with the mouth cone directed posteriorly. Generally the antennae, which in most species are nine-segmented, are frequently reduced by fusion and project forward in front of the eyes. Of the mouthparts only the single left mandible is developed and the maxillae are modified into piercing stylets. Both the maxillae and the labium bear segmented sensory projections (palps).

The middle portion of the body (thorax) bears the legs. The first part of the thorax (prothorax) of thrips bears the forelegs, the inner parts (coxae) of which may exhibit ridges that when rubbed with spurs from the lower parts (femurs) emit sounds, although these cannot be detected by the human ear without amplification. The remainder of the thorax (pterothorax) bears the wings as well as the mid and hind legs. The wings are usually fairly simple and straplike. At rest they are laid over the abdomen but never folded. Typically the wings have long posterior hairlike fringes with shorter fringes occurring on the leading edge.

The abdomen is elongated and generally flattened dorso-ventrally, especially in the Tubulifera. There are 10 distinctive segments with a rudiment of the 11th. The first and eighth abdominal segments bear breathing holes (spiracles). Often in the Tubulifera, several setae on the back are S-shaped and hook into the wing fringes to hold the wings at rest. In the Terebrantia the terminal segment is rounded in the males and split ventrally up to the eighth segment in the females. In the Tubulifera the terminal segment is tubelike.

The male genital organs are composed of paired appendages and an unpaired copulatory organ (aedeagus), all of which retract into the abdomen. The paired testes and accessory glands with their ducts occur in the posterior half of the abdomen. The female of the Terebrantia usually has an ovipositor composed of two pairs of sawlike blades. These are absent in the Tubulifera. There are eight egg sacs (ovarioles) and spherical, often pigmented, seminal receptacle for storage of sperm acquired during mating.

Management

Thrips are difficult to control. If management is necessary, use an integrated program that combines the use of good cultural practices, natural enemies, and the most selective or least-toxic insecticides that are effective in that situation.

Monitoring

If thrips are a suspected cause of plant damage, thrips adults and larvae can be monitored by branch beating or gently shaking foliage or flowers onto a light-colored sheet of paper, beating tray, or small cloth. For thrips that feed in buds or unexpanded shoot tips, clip off several plant parts suspected of harboring thrips, place them in a jar with 70% alcohol (ethanol), and shake vigorously to dislodge the thrips. Strain the solution through filter paper so thrips can more readily be seen.

Adult thrips can also be monitored by hanging bright yellow sticky traps in or near host plants.

Be aware that the presence of thrips does not mean that damage will result from their feeding. Large numbers of thrips in traps, or adults in flowers feeding on pollen, do not necessarily indicate that control action is needed. Plants suspected of being infected by thrips-vectored viruses can be reliably diagnosed only by sending properly collected samples from symptomatic plants to a laboratory that tests for plant pathogens.

Biological Control

Predatory thrips, green lacewings, minute pirate bugs, mites, and certain parasitic wasps help to control plant-feeding thrips. To conserve and encourage naturally occurring populations of these beneficials, avoid creating dust and consider periodically rinsing dust off of small plants, avoid persistent pesticides, and grow a diversity of plant species.

Where thrips are a problem, learn whether that pest has specific natural enemies important in its control. For example, a minute pirate bug, *Macrotracheliella nigra*, and green lacewing larvae are important predators of Cuban laurel thrips. *Euseius* species mites are important predators of citrus thrips. With greenhouse thrips in Southern California up to 50% of its eggs are killed by a tiny wasp, *Megaphragma mymaripenne*. After feeding inside during its larval stage then pupating, the emerging adult parasite leaves a relatively large round hole in the tiny thrips egg. Conversely, when a greenhouse thrips emerges from an unparasitized egg, part of the egg shell is often visible at the side of the egg blister. *Thripobius semiluteus* parasitizes greenhouse thrips larvae. Thrips parasitized by this wasp's larvae become swollen around the head and turn black, in contrast to the pale color of unparasitized greenhouse thrips larvae. Unlike healthy black mature thrips, the black parasitized larvae are smaller and do not move.

There is little research-based information on the effectiveness of releasing thrips natural enemies in gardens and landscapes. Releasing purchased natural enemies, in most situations, is unlikely to provide satisfactory thrips control.

Cultural Control

Thrips species that feed on many different plant species often move into gardens and landscapes when plants in weedy areas or grasslands begin to dry in spring or summer. Avoid planting susceptible plants next to these areas, and control nearby weeds that are alternate hosts of pest thrips. Grow plants that are well-adapted to conditions at that site. For example, plants adapted to grow in full sun can be stressed when planted in shady conditions and may be more susceptible to thrips damage. Provide appropriate cultural care to keep plants vigorous and increase their tolerance to thrips damage. Keep plants well irrigated, and avoid excessive applications of nitrogen fertilizer, which may promote higher populations of thrips. Old, spent flowers can harbor thrips, so their removal and disposal is sometimes recommended. However, the general benefit of this practice in landscapes is unknown; and old blossoms also commonly shelter beneficial predators of thrips.

Investigate the availability of resistant cultivars. For example, western flower thrips more often damages fragrant, light-colored, or white roses. Rose cultivars, with sepals that remain tightly wrapped around the bud until just before blooms open, have fewer thrips problems. Where Cuban laurel thrips is a problem on Indian laurel fig you can plant *Ficus microcarpa*, "Green Gem," which is mostly resistant to this pest.

Pruning

Prune and destroy injured and infested terminals when managing a few small specimen plants in the landscape. Avoid shearing plants, which is the clipping of dense foliage to maintain an even surface on formal hedges or creating specific shapes (topiary). Shearing stimulates thrips-susceptible new growth. Prune by cutting plants just above branch crotches and nodes instead of shearing off terminals.

Prune during specific times of the year to help control certain thrips. Prune off galled, rolled terminals of Indian laurel fig during winter to greatly reduce thrips damage the next summer. Relatively few Cuban laurel thrips can survive the winter outside of the protection provided by the leaves they gall. Instead of pruning avocado during February through April, January pruning may reduce thrips scarring of fruit. January pruning can induce additional avocado growth flush during May fruit set and reduce thrips scarring of fruit. When succulent foliage is abundant in spring thrips tend to remain and feed on leaves and not move to fruit. Pruning the interior of citrus trees can increase predaceous mite populations in the exterior canopy, thereby reducing fruit scarring by citrus thrips.

Row Covers

Row covers, hot caps, and other types of cages with a fine mesh can exclude thrips and other insects from vegetables and other young herbaceous plants. Apply row covers before crops emerge or to pest-free plants during planting. Plants are normally covered or caged only while they are young and most susceptible to damage. Once plants become larger or temperatures get warmer, remove covers to provide enough growing space and to prevent overheating. Drip or furrow irrigation is generally necessary when using row covers.

Any type of covering that excludes insects but allows light and air penetration can be used. With sturdy crops that do not grow too tall, floating row covers (vented polyethylene, spunbonded polyester, point-bonded polypropylene) can be placed on top of beds with no frames or hoops. The crop itself lifts the fabric as it grows. For plants that grow upright or have sensitive tips that might be damaged when pushing against covers, use hoops, plastic tunnels, or wire strung between posts to hold up covers. Wood, wire, or plastic frames covered with muslin, nylon, or other fine mesh can be used for several years.

Reflective Mulch

Mulch or mesh that reflects light interferes with certain flying insects' ability to locate plants. If the plants are initially pest-free, and relatively small in comparison with the surface area that is covered with reflective material, reflective mulch can delay or reduce the extent to which young plants become infested by winged aphids and adult leafhoppers, thrips, and whiteflies. In flower and vegetable crops that are especially sensitive to insect-vectored viruses, the cost and effort of using reflective mulch may be justified; because the mulch can be significantly more effective than insecticides in preventing or delaying infection of small plants. As plants grow larger, reflective mulch becomes increasingly less effective and other management methods may be needed. Reflective mulch ceases to repel insects by the time the plant canopy covers more than about half of the soil surface.

Silver or gray is the most effective color for synthetic reflective mulch or mesh, but white also works. Some organic mulches (e.g., straw) and living mulches (e.g., interplanting the crop with buckwheat, *Fagopyrum esculentum*) may also repel certain pests, but this is not as well documented. Commercially available synthetics include aluminum-metalized polyethylene and silver-embossed polyethylene plastic films. If handled carefully, these may be used for more than one season. Aluminum foil is also effective and may be suitable for a small garden; but it is expensive and difficult to reuse because it is delicate to handle.

Synthetic mulch application methods include:

- Transplant seedlings through holes in the mulch.
- Apply the mulch before plants emerge from the soil and leave a narrow, mulch-free strip along the planting row.
- Lay lightweight material that allows light and air penetration over the top of a crop that is sturdy enough to lift the material as it grows.

In addition to temporarily repelling certain flying insects, mulch may improve growth of certain crops by increasing light levels, keeping soil warmer overnight, reducing weed growth, and conserving soil moisture. Mulch can also have negative effects. It can increase crop susceptibility to root diseases, prevent the use of overhead watering, and make it more difficult to know whether you are maintaining proper soil moisture in the rooting zone of a plant. Most recyclers will not accept plastics with soil on them; therefore, plastic mulches typically are disposed of in landfills. Investigate which material and methods are likely to work best in your situation.

Chemical Control

Although thrips damage is unsightly, it does not usually warrant the use of insecticides in gardens and landscapes. Feeding injury typically does not become apparent until after tissue grows and expands. Thus, by the time damage is noticed on ripening fruit or distorted terminals, the thrips that caused the damage are often gone. No pesticide application will restore the appearance of injured tissue; plants will remain damaged until leaves drop, injury is pruned off, or new unblemished fruit is produced. Where plant viruses are a problem, insecticides typically do not kill thrips fast enough to prevent the transfer of virus from thrips to plants. Using row covers or other methods to prevent thrips infestation is the most effective way to prevent infection by thrips-vectored viruses.

Thrips can be difficult to control effectively with insecticides, partly because of their mobility, feeding behavior, and protected egg and pupal stages. Improper timing of application, failure to treat the proper plant parts, and inadequate spray coverage when using contact materials are common mistakes that can prevent potentially effective insecticides from actually providing control. Before using a pesticide, learn more about the biology of your pest species and the characteristics of available products by reading the label and consulting the Active Ingredients Database. Often you will learn chemical control cannot be effective until the next season, when new plant growth develops. Certain products are available only by hiring a professional applicator. If insecticides are used, combining their use with appropriate cultural practices and other methods usually improves the pest control.

Insecticides most Compatible with IPM

Contact insecticides that do not leave persistent residues can be effective for greenhouse thrips and other species that feed openly on plants. These products have low toxicity to people, pets, and pollinators and relatively little adverse impact on biological pest control; because they do not leave toxic residues that would kill natural enemies migrating in after their application. Contact insecticides include azadirachtin, insecticidal soaps, narrow-range oil, neem oil, and pyrethrins, which many products combine with piperonyl butoxide. To be effective, contact sprays must be applied to thoroughly cover buds, shoot tips, and other susceptible plant parts where thrips are present. On plants with a history of unacceptable damage, begin treatment early when thrips or their damage is first observed. Unless directed otherwise by the product label, periodically repeat the application as long as pest thrips and susceptible plant parts are both present.

Spinosad is generally more effective against thrips than the products above. Spinosad lasts 1 week or more and moves short distances into sprayed tissue (has translaminar activity) to reach thrips feeding in protected plant parts. Adding horticultural oil to the spray mix can increase its persistence within plant tissue. This insecticide is a fermentation product of a naturally occurring bacterium, and certain formulations are organically acceptable. Spinosad can be toxic to certain natural enemies (e.g., predatory mites, syrphid fly larvae) and bees when sprayed and for about 1 day afterward; do not apply spinosad to plants that are flowering.

Insecticides More Toxic to Thrips and Beneficial Insects

Systemic insecticides are absorbed by one plant part (e.g., roots) and moved (translocated) to other plant parts. Trunk spray or injection of an effective, systemic, neonicotinoid insecticide can provide relatively rapid control. With soil drench or injection, there is a longer time delay between neonicotinoid application and insecticide action. Neonicotinoids vary in effectiveness for thrips control. For example, dinotefuran, available to professional applicators, can provide good control of thrips. Imidacloprid commonly fails to provide satisfactory thrips control, and imidacloprid generally is not recommended for thrips.

Neonicotinoids have low, moderate, or severe adverse impact on natural enemies and pollinators varying with the product, situation, and the species and life stage of invertebrate. Neonicotinoid insecticides can translocate to flowers and may harm natural enemies and pollinators that feed on nectar and pollen. Delay systemic insecticide application until after plants have completed their seasonal flowering, unless the product label directs otherwise. Where the roots of nearby plants grow near treated plants, those other plants may also take up some of the soil-applied insecticide. Delay soil application, when possible, until after the nearby plants are also done flowering.

To avoid tree injury and potential spread of pathogens on contaminated tools, use a soil application or trunk spray whenever possible, instead of injecting or implanting trees with insecticide. Injecting or implanting trunks or roots injures trees, and it is difficult to repeatedly place insecticide at the proper depth. If injecting or implanting multiple trees, prevent the potential spread of pathogens on contaminated tools; before moving to work on each new tree, scrub any plant sap from tools or equipment that penetrate trees and disinfect them with a registered disinfectant (e.g., bleach). Avoid methods that cause large wounds, such as implants placed in holes drilled in trunks. Do not implant or inject roots or trunks more than once a year.

Insecticides to Avoid

The systemic organophosphate acephate is available for ornamental, nonfood plants. Avoid using it. Acephate can be highly toxic to natural enemies and pollinators and can cause spider mites to become abundant and damage plants after its application.

Avoid foliar sprays of other organophosphate insecticides (e.g., malathion), carbamates (carbaryl), or pyrethroids (e.g., bifenthrin, cyfluthrin, fluvalinate, and permethrin). These materials are highly toxic to natural enemies and pollinators, can cause spider mite outbreaks, and are not particularly effective against most thrips. Because their use in landscapes and gardens can run or wash off into storm drains and contaminate municipal wastewater, these insecticides are being found in surface water and are adversely affecting nontarget, aquatic organisms.

Greenhouse Thrips Management

Greenhouse thrips can infest many plant species but primarily is a pest of evergreen, broadleaved perennials. It occurs mainly on the underside of leaves and on fruit clusters or other plant parts that touch each other. Greenhouse thrips is sluggish and the adults tend not to fly. Individuals feed in groups and populations usually begin in a limited part of the plant and spread slowly. If the underside of leaves on susceptible plants are regularly inspected to allow early detection and removal of new infestations, pruning off colonies can be effective.

Greenhouse thrips is readily controlled with thorough application of contact sprays such as horticultural oil, natural pyrethrins (plus piperonyl butoxide), or insecticidal soaps to the underside of infested leaves. Repeat applications may be necessary. Keep in mind that greenhouse thrips have natural enemies in the landscape (discussed above). Assess whether spraying is warranted and select materials that are least toxic to natural enemies.

Table: Some Common Pest Thrips and Their Host Plants.

Thrips			**Appearance**[1]		
Common Name	**Scientific Name**	**Host Plants**	**Primary Damage**	**Adults**	**Larval Body**
avocado thrips	*Scirtothrips perseae*	avocado	scabby brown scars on fruit	3 red spots atop head, brown lines separating segments on pale yellow abdomen	pale yellow
bean thrips	*Caliothrips fasciatus*	bean, occasionally other legumes	brown, distorted leaf and seedling terminals	blackish body with white wing bands	yellow to orangish
citrus thrips	*Scirtothrips citri*	blueberries and citrus, generally not damaging to its many other hosts	scabby silvery scars on citrus fruit; distorted blueberry leaves and shoots	light orangish yellow to white body	light orangish yellow to white
Cuban laurel thrips	*Gynaikothrips ficorum*	laurel fig or Indian laurel, *Ficus microcarpa*	rolled, podlike, darkscarred terminals; galls	blackish body	yellowish to white
greenhouse thrips	*Heliothrips haemorrhoidalis*	mostly perennials with thick, broad leaves, including avocado, azalea, hypericum, laurel (English and Grecian), photinia, and rhododendron	leaves bleached with black excrement on undersides; scabby fruit	black body with pale wings	white to yellowish

myo-porum thrips	*Klambothrips myopori*	*Myoporum laetum, M. pacificum*	leaves swollen, curled, and distorted; terminals galled	black body	early instars are white to yellowish, pupae are orangish
onion thrips	*Thrips tabaci*	vegetables including garlic, onion, and pepper; many herbaceous ornamentals where it's usually not damaging	stippled and scarred petals, leaves, and other plant parts; distorted terminals	yellow to dark brown body	yellow to orangish
toyon thrips	*Liothrips ilex*	Christmas berry or toyon	crinkled, undersized, sometimes blackened terminal leaves	black body with pale wings	yellow
western flower thrips	*Frankliniella occidentalis*	many herbaceous ornamentals (impatiens, petunia); vegetables (cucurbits, pepper); fruits (grape, strawberry); some shrubs and trees (rose, stone fruit)	stippled and scarred petals, leaves, and other plant parts; distorted terminals, vectors tospoviruses affecting herbaceous plants	thick, bristle-like hairs at the tip of the abdomen; abdomen extends beyond wing tips at rest; individuals vary greatly in color	yellow to orangish

[1] Color does not reliably distinguish among thrips, which can be accurately identified to species only by an expert examination of microscopic characters.

Table: Some Common Beneficial Predatory Thrips.

Thrips		**Appearance**		
Common Name	**Scientific Name**	**Where observed**	**Adults**	**Larvae**
banded-wing thrips	*Aeolothrips*spp.[1]	among pest mites and thrips	black body, white wings with black bands	yellow body
black hunter thrips	*Haplothrips mali*[2]	among mites, scales, and pest thrips	dark brown or entirely black body with long abdomen, white wings, much more active than similar-looking greenhouse thrips	dark, reddish-brown body

Franklinothrips or vespiform thrips	*Franklinothrips orizabensis, F. vespiformis* [1]	among lace bugs, mites, and pest thrips	mostly black body, with pale or white areas; distinctly narrow where abdomen meets thorax	yellow to orange body, swollen abdomen with red or dark orange band, body more stout or oval-shaped than most thrips
sixspotted thrips	*Scolothrips sexmaculatus* [3]	in colonies of mites	3 dark blotches on each forewing, body pale to yellowish	yellow to whitish body
Predatory thrips can sometimes be distinguished from pest species because predators are seldom seen at high levels as can be common with certain pest thrips.				

Families: [1] Aeolothripidae; [2] Phlaeothripidae; [3] Thripidae

Cockchafer

This is the Common cockchafer, also referred to as the May bug, the Spang beetle or the Billy witch. These beetles usually appear around late April – early May and can frequently be seen and heard flying into lit windows and even lamps indoors.

Common cockchafer males can easily be distinguished from the females by counting the number of 'leaves' on their remarkable antler-like antennae, males sport seven 'leaves' while females have only six. These leafy antennae can detect pheromones, enabling males to find females even in the dark.

Lifecycle

Life begins as an egg laid around June - July, hatching into a white grub which lives underground. Grubs can spend 3 years underground (up to 5 years in colder climates) until they pupate. As grubs they munch on roots and tubers until they reach around 4cm. This is the point when they pupate, emerging as an adult beetle (or imago) in the spring. They live as adults for a mere six weeks during which time the female can lay as many as 80 eggs.

Cockchafers were once highly abundant until pesticide use in the mid 20th Century almost obliterated them. Thankfully they have been making a come-back since the 1980's with the regulation of pesticides.

The larvae are considered to be an agricultural nuisance since they can be highly destructive to crops. Before agricultural intensification these beetles were especially problematic, so much so that adults were caught and killed to break the life cycle, in 1911, more than 20 million individuals were collected in 18 km^2 of forest. A less conventional approach was taken during 1320, when the cockchafers (as a species) were taken to court in Avignon where they were ordered to leave town and relocate to a specially designated area, or be outlawed. All cockchafers who failed to comply were collected and killed. Both adults and grubs have been considered a delicacy at times and are still eaten in some countries.

In ancient Greece, young boys used to catch the unwitting cockchafer, and tether it by tying a thread around its feet, amusing themselves by watch the poor chap fly aimlessly around in spirals.

Vertebrate Pests

Vertebrate pests include birds, mammals, or reptiles that cause damage to agricultural crops. A number of introduced animals – including rabbits, feral cats, foxes, house mice, wild dogs, Indian Myna, and pigs – have established large and widespread populations in Australia and are pests to crops and livestock.

Types and Damage caused

Rodents

Include Indian palm squirrel; long-haired or plague rat; and mice. Significant damage is caused by mice plagues. They attack all cereal and grain crops, as well as many vegetables and fruits.

Native and Exotic Animal Pests

Include bats; possums; and ferrets. Commonly damaged vegetable crops include broccoli, tomatoes, capsicum, celery, lettuce.

Feral Cats

These are domestic cats which survive and reproduce without close association to humans. However, their status is variable and feral cats may live either in close proximity to human habitation or be completely isolated in the bush.

Foxes

Or the red foxes are pest animals and known to kill livestock, particularly the young, and also damage vegetable and fruit crops.

Rabbits

Feral or wild populations of rabbits selectively feed on certain species of plants at critical stages of development such as seeding and seedling establishment. Vegetables often targeted include beans, peas, beet, broccoli, carrot, lettuce, and herbs like parsley.

Feral Pigs

Feral pigs are the descendants of domestic pigs that escaped and have bred in the wild. They spread widely and rapidly throughout the northern and eastern regions of the country, creating considerable amounts of damage wherever they became established. They rest and shelter in crops, eat crops during harvest time, trample pasture, and cause serious soil disturbance when they uproot large tracts of ground – often new crops – in search of roots and young plant shoots.

Wild Dogs

Wild dogs can be a major pest threatening livestock and production in Australia.

Birds

Include cockatoos; other parrots; and Indian Myna. Pest bird problems are increasing in Australia, particularly with recent expansions in the grape and wine industry, together with the olive industry. More than 20 species of birds conflict with primary production by significantly reducing the profitability of a wide range of crops in horticultural industries.

Pest Management

There are a range of methods available to control feral animals.

- Methods of pest bird control include non-lethal techniques such as scaring devices, chemical repellents, habitat manipulation, use of decoy food sources, and exclusion netting.
- Feral cats are widespread throughout Australia, and therefore, no large-scale control methods can be applied to them. However, traditional control measures such as shooting and trapping can be used.
- Baiting, trapping, shooting, exclusion fencing, and adoption of good animal husbandry practices are some of the methods to manage and control wild dogs.
- Fox populations are very resilient to conventional methods of control. Rapid re-invasion of areas occurs after control measures are applied. A control program must ensure that all of the species is at risk from the control measures chosen. Effective control integrates a knockdown effect from large-scale poisoning programs combined with den fumigation; exclusion fencing; strategic shooting; and appropriate animal husbandry.
- An integrated rabbit control (IRC) program is the combination of several measures used to achieve long-term results which is used to minimise the rabbit impact. Poisoning; removal of above ground surface harbour (fallen timber, log heaps, rocks, introduced weeds, briars, and discarded building/fencing materials); warren destruction (rabbits need burrows to breed and survive predators); fumigation; night shooting; and trapping.

Rodent

Rodent, (order Rodentia), any of more than 2,050 living species of mammals characterized by upper and lower pairs of ever-growing rootless incisor teeth. Rodents are the largest group of mammals, constituting almost half the class Mammalia's approximately 4,660 species. They are indigenous to every land area except Antarctica, New Zealand, and a few Arctic and other oceanic islands, although some species have been introduced even to those places through their association with humans. This huge order of animals encompasses 27 separate families, including not only the "true" rats and mice(family Muridae) but also such diverse groups as porcupines, beavers, squirrels, marmots, pocket gophers, and chinchillas.

General Features

All rodents possess constantly growing rootless incisors that have a hard enamel layer on the front of each tooth and softer dentine behind. The differential wear from gnawing creates perpetually sharp chisel edges. Rodents' absence of other incisors and canine teeth results in a gap, or diastema, between incisors and cheek teeth, which number from 22 (5 on each side of the upper and lower jaws) to 4, may be

rooted or rootless and ever-growing, and may be low- or high-crowned. The nature of the jaw articulation ensures that incisors do not meet when food is chewed and that upper and lower cheek teeth (premolars and molars) do not make contact while the animal gnaws. Powerful and intricately divided masseter muscles, attached to jaw and skull in different arrangements, provide most of the power for chewing and gnawing.

Capybara (*Hydrochoerus hydrochaeris*)

The range in body size between the mouse (18 grams [0.64 ounce], body 12 cm [4.7 inches] long) and the marmot (3,000 grams, body 50 cm long) spans the majority of living rodents, but the extremes are remarkable. One of the smallest is Delany's swamp mouse (*Delanymys brooksi*), associated with bamboo in the marshes and mountain forests in Africa. It weighs 5 to 7 grams, and the body is 5 to 6 cm long. The largest is the capybara (*Hydrochoerus hydrochaeris*) of Central and South America, which weighs 35 to 66 kg (77 to 146 pounds) and stands 50 to 60 cm at the shoulder, with a body 100 to 135 cm long. Some extinct species were even larger, attaining the size of a black bear or small rhinoceros. The largest rodent ever recorded, *Josephoartigasia monesi*, lived some two to four million years ago, during the Pleistocene and Pliocene epochs; by some estimates it grew to a length of about 3 metres (10 feet) and weighed nearly 1,000 kg.

Importance to Humans

Rodents have lived on the planet for at least 56 million years and modern humans for less than one million, but the consequences of their interactions during that short overlap of evolutionary time have been profound. For rodents, early humans were just another predator to avoid, but with *Homo sapiens'* transition from nomadic hunting and gathering to sedentary agricultural practices, humans became a reliable source of shelter and food for those species having the innate genetic and behavioral abilities to adapt to man-made habitats. The impact of these species upon human populations ranges from inconvenient to deadly. Crops are damaged before harvest; stored food is contaminated by rodent waste; water-impounding structures leak from burrowing; and objects are damaged by gnawing. Certain species are reservoirs for diseases such as plague, murine typhus, scrub typhus, tularemia, rat-bite fever, Rocky Mountain spotted fever, and Lassa fever, among others. Only a few species are serious pests or vectors of disease, but it is these rodents that are most closely associated with people.

Various other rodents are beneficial, providing a source of food through hunting and husbandry, apparel derived from their fur, test animals for biomedical and genetic research (especially mice and rats), pleasure as household pets, and insight on mammalian biology and evolutionary history.

As Pests and Disease Vectors

Rodents cause significant losses to crops, such as these potatoes damaged by voles.

Some rodent species are serious agricultural pests, eating large quantities of food stored by humans. For example, in 2003, the amount of rice lost to mice and rats in Asia was estimated to be enough to feed 200 million people. Most of the damage worldwide is caused by a relatively small number of species, chiefly rats and mice. In Indonesia and Tanzania, rodents reduce crop yields by around fifteen percent, while in some instances in South America losses have reached ninety percent. Across Africa, rodents including *Mastomys* and *Arvicanthis* damage cereals, groundnuts, vegetables and cacao. In Asia, rats, mice and species such as *Microtus brandti*, *Meriones unguiculatus* and *Eospalax baileyi* damage crops of rice, sorghum, tubers, vegetables and nuts. In Europe, as well as rats and mice, species of *Apodemus*, *Microtus* and in occasional outbreaks *Arvicola terrestris* cause damage to orchards, vegetables and pasture as well as cereals. In South America, a wider range of rodent species, such as *Holochilus*, *Akodon*, *Calomys*, *Oligoryzomys*, *Phyllotis*, *Sigmodon* and *Zygodontomys*, damage many crops including sugar cane, fruits, vegetables, and tubers.

Rodents are also significant vectors of disease. The black rat, with the fleas that it carries, plays a primary role in spreading the bacterium *Yersinia pestis* responsible for bubonic plague, and carries the organisms responsible for typhus, Weil's disease, toxoplasmosis and trichinosis. A number of rodents carry hantaviruses, including the Puumala, Dobrava and Saaremaa viruses, which can infect humans. Rodents also help to transmit diseases including babesiosis, cutaneous leishmaniasis, human granulocytic anaplasmosis, Lyme disease, Omsk hemorrhagic fever,

Powassan virus, rickettsialpox, relapsing fever, Rocky Mountain spotted fever, and West Nile virus.

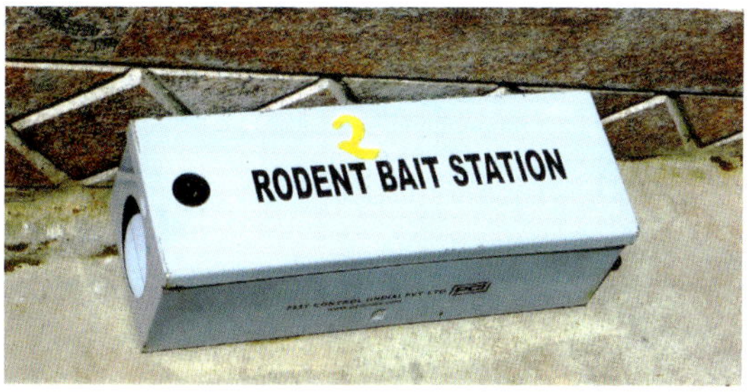

Rodent Bait Station, Chennai, India

Because rodents are a nuisance and endanger public health, human societies often attempt to control them. Traditionally, this involved poisoning and trapping, methods that were not always safe or effective. More recently, integrated pest management attempts to improve control with a combination of surveys to determine the size and distribution of the pest population, the establishment of tolerance limits (levels of pest activity at which to intervene), interventions, and evaluation of effectiveness based on repeated surveys. Interventions may include education, making and applying laws and regulations, modifying the habitat, changing farming practices, and biological control using pathogens or predators, as well as poisoning and trapping. The use of pathogens such as *Salmonella* has the drawback that they can infect man and domestic animals, and rodents often become resistant. The use of predators including ferrets, mongooses and monitor lizards has been found unsatisfactory. Domestic and feral cats are able to control rodents effectively, provided the rodent population is not too large. In the UK, two species in particular, the house mouse and the brown rat, are actively controlled to limit damage in growing crops, loss and contamination of stored crops and structural damage to facilities, as well as to comply with the law.

Rodent Control and Prevention

1. Direct disinfestations
   - Use poisonous bait or traps
   - Temporary effect

   a. Stomach poison
      - Mix rodenticide with bait for rodent consumption
      - The rodenticides used by our department are chronic anti-coagulants (vitamin K as the anti-dote)
      - The anti-coagulants take 4-5 days to give effects

b. Traps

- Cage trap
- Snap trap/break back trap
- the use of glue traps can scare rodent that are caught live and cause them to urinate. Since their urine may contain germs, this may increase the risk of being exposed to rodent-borne diseases

2. Fundamental control

Improve the sanitary condition of the environment and deprive rodents of :

- Food
- Harbourage
- Passages

Purposes of Rodent Prevention

- Eliminate harbourages for rodents and confine the rodent activity area for enhancing the rodent disinfestations work
- Prevent re-infestation of rodent

Disposal of Dead Rodents

The following procedures can be used for handling dead rodents found:

- using tools such as tongs to put the dead rodents into a tough plastic bag (e.g. rubbish bag);
- spraying the carcasses with general household disinfectant or diluted bleaching solution until they are soaked thoroughly;
- sealing the bag tightly and placed it into another plastic bag. The second plastic bag should also be sealed;
- putting the bagged material properly in covered rubbish bin or disposing of it to the nearest refuse collection point.

When handling dead rodents, attention should be paid to personal and environmental hygiene. Wear gloves and face mask, if necessary, when handling dead rodents and avoid direct contact with them. All areas, clothes and items contaminated by the dead rodents should be disinfected thoroughly using general household disinfectant or diluted bleaching solution. Before taking off gloves, wash them with water and then cleanse with general household disinfectant or diluted bleaching solution. Hands should be washed thoroughly with soap and water after removing the gloves.

Rattus norvegicus, body length (head to tail) 34-46cm, body weight 150-600g

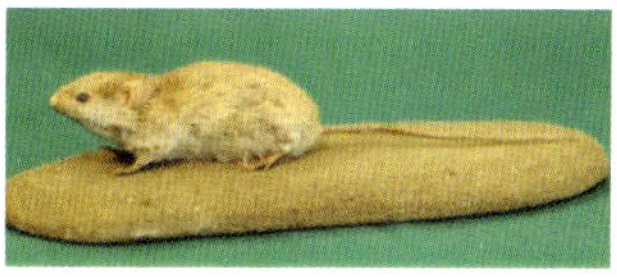

Mus musculus, body length (head to tail) 12-21cm, body weight 10-21g

R. sanguineus, a tick, is a rodent-borne disease vector for the transmission of Boutonneuse fever

Rattus rattus, body length (head to tail) 35-46cm, body weight 80-300g

Ctenocephalides felis, about 2-4mm long

L. deliense, a mite, is a rodent-borne disease vector for the transmission of scrub typhus

The Norway rat, roof rat, and house mouse are destructive rodent pests in and around farm facilities. This can be especially true during the winter months, as they seek food and refuge indoors. Rats and mice consume and contaminate feed, gnaw on structural, mechanical, and electrical components, and weaken concrete slabs and walkways with their burrowing activities. They can also potentially carry diseases such as bubonic plague, leptospirosis, rabies, and bacterial food poisoning.

Usually the first signs of rodent infestation are droppings or urine stains in and around buildings, because rats and mice are most active at night. If rodents are seen repeatedly during the day, it is an indication of an established population. It is estimated that for every rodent seen during the day around barns and poultry houses, there are likely 20 to 50 that are unseen.

Effective rodent control involves a three step process. The first step is to "rodent-proof" the structure. This is very difficult because rats and mice can squeeze through holes just large enough to pass their heads through, as small as ¼ for mice and ½ inch for rats. They can climb through pipes, jump vertically three feet, horizontally four feet, and climb wires, cables, vines, and trees to enter a building.

To be effective, rodent-proofing must block all possible entry points. Openings must be sealed with mortar, concrete, sheet metal, or hardware cloth (19 gauge or heavier with no openings more than ¼ inch) around augers, pipes and wires where they enter structures. Also, corrugated metal siding should be sealed and corner seams made tight.

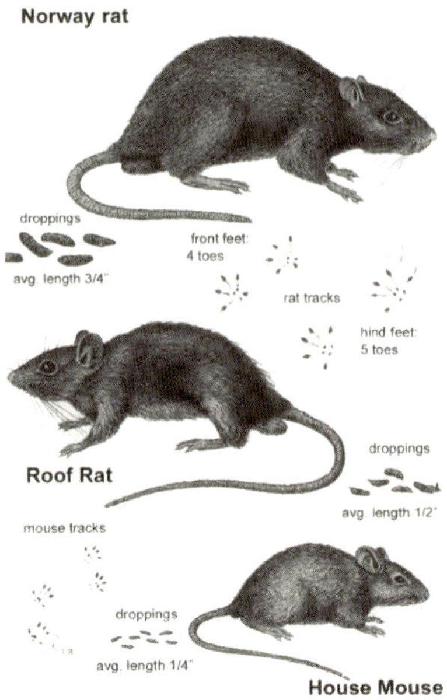

Controlling Rodents in Commercial Poultry Facilities Purdue Extension

Good sanitation is step number two. Food, water, and nesting material must be eliminated. Recommended practices include:

- Clean up debris and trash.
- Store feed in metal cans with tight fitting lids.
- Keep grass short and maintain at least a 3-foot space around the building that is free of brush, trash, and weeds.
- Dry up water sources such as dripping faucets or leaking pipes.

The third step is population reduction with traps and baits. Since rodents have a small home range, rats travel no more than 100 feet and mice less than 30 feet from their nesting site, trapping is an effective, quick and economical method of control. Trapping is often underrated, especially where only a few rodents are present. Common snap traps, glue boards, and live traps can be used to supplement baiting programs, or in situations where baits may pose a hazard. Traps should be placed along walls, near holes, or at right angles along beams, rafters, or other travel ways. Traps may be baited with a variety of food items such as whole nuts, peanut butter, or small pieces of meat.

There are many different types of rodenticides (poison baits) on the market. They may be formulated as bar baits, pellets, concentrates, or tracking powders (table below). In most situations, ready-to-use commercial baits are preferred over mix-your-own baits, because they do not require the applicator to handle a concentrated toxicant. Bar baits are formulated with a high wax content for outdoor use and high humidity areas. Pellets are formulated with grain and a binder that holds the pellet together for indoor use. Concentrates and tracking powders are occasionally used by professional pest control operators.

Janet Hurley~ Texas A&M Agrilife Extension Service~ Bugwood.org

Rodenticides are classified into two broad chemical groups: anticoagulants and non-anticoagulants. The older, or first generation, anticoagulants were first discovered in the 1940s. Warfarin was the first on the market and became the best known and most widely used. However, its use is limited today due to resistance and newer, more potent anticoagulants that are available. Most of the first generation anticoagulants are multiple-dose baits: that is they cause death only after they are eaten several days in a row. The newer second generation anticoagulants are effective after a single dose. These include brodifacoum, bromadiolone, and difethialone. The single-dose anticoagulants are generally effective against rodents resistant to the older multi-dose compounds.

The two most common non-anticoagulant baits are bromethalin and cholecalciferol or vitamin D^3. Bromethalin kills rodents by shutting down their ability to produce energy within the cells of the body. A single-dose is usually lethal within 1 to 3 days. It is considered safe, and a minor threat of secondary poisoning, because it causes the rodent to stop feeding days before it dies, so most of the poison has been excreted prior to death and possible ingestion by a predator or pet. Cholecalciferol is a calcium releaser that causes too much calcium to be released into the blood, disrupting body functions. Cholecalciferol kills anticoagulant resistant rodents and there is no problem of secondary poisoning of pets or wildlife that eat poisoned rodents. Cholecalciferol will act as a single-dose poison, if a sufficient amount is consumed by a rodent in one feeding, but it will act as a multiple-dose poison if consumed in lesser amounts over several days.

Zinc phosphide is a single-dose bait which has been used for many years. This is a restricted-use product, which is mostly used for agricultural rodent control, because it offers little risk of secondary poisoning of beneficial predators. However, it has no antidote and is not appropriate for use around children, pets, or livestock.

When using rodenticides, safety must be the first consideration. Poison baits must be placed where they are inaccessible to children, pets, livestock, and wildlife. Tamper-proof bait stations should be used where rodent runs are exposed and in all outdoor situations.

Table: Common Rodenticide Baits				
Active Ingredient	Trade Names	Type	Formulation	Dose Required
ANTICOAGULANTS				
chlorophacinone	Rozol, Ramucide, Microzul, other brands	1st generation	Tracking powder, pellets, blocks, concentrate	Multiple dose; death 5-7 days after feeding.
diphacinone	Ramik Green, Contrax-D, Ditrac, Trap-N-Sak, other brands	1st generation	Pellets, place packs, bait blocks	Multiple dose; death 5-7 days after feeding.
pindone	Pival, Pivalyn	1st generation	Bait packs and blocks, bulk	Multiple dose; death 5-10 days after feeding.
warfarin	Ferret, Contrax, other brands	1st generation	Place packs, bulk	Multiple dose; death 5-10 days after feeding.
brodifacoum	Attack, Havoc, Jaguar, D-Con, Talon, other brands	2nd generation	Place packs, bulk, pellets, bait blocks	Single dose; death 5-7 days after feeding.
bromadiolone	Boot Hill, Contrac, Hawk, Just One Bite, Maki, Tomcat 2, other brands	2nd generation	Place packs, bulk pellets, bait blocks	Single dose; death 5-7 days after feeding.
difethialone	Generation, Hombre	2nd generation	Pellets, place packs, bait blocks	Single dose; death 5-7 days after feeding.
NON-ANTICOAGULANTS				
bromethalin	Assualt, Fastrac, Vengeance, Tomcat, Trounce, other brands	Neurotoxin	Place packs, bulk, bait blocks	Single dose; death 2-4 days after feeding.
cholecalciferol	Rampage, Quintox, Terad3 Blox, other brands	Vitamin D3 calcium releaser	Place packs, bait blocks	Single or multiple dose; death 3-5 days after feeding.
Zinc phosphide "Restricted-Use Pesticide"	Many brands	Stomach poisoning from phosphine gas formation.	Pellets, mixed grains	Single dose (acute); quick knock down.

Common Myna

The common myna is a native of India, east and west Pakistan and Burma. It was introduced to many Pacific lands, including New Zealand, usually to combat invertebrate pests. Mynas are large, conspicuous passerines. A shiny black head and shoulder plumage merges into vinous brown for the remainder of the body and a large patch of white is flashed from each wing during flight. They are commonly seen deftly avoiding traffic while foraging for road-killed insects. At dusk small groups join communal roosts, which may comprise many hundreds of individuals. Mynas, originally widely introduced to New Zealand in the 1870s, are found in all suitable manmade habitat north of 40° S, including forest remnants and many offshore islands, but are in low numbers or absent from some parts of the volcanic plateau and large forest tracts such as Urewera.

Identification

Mynas are stocky brown birds with a shiny black head and shoulders. They are commonly seen in pairs or in small family groups. Adult primaries are black when new but fade to brown. The base of primaries is white and all upper and lower coverts are also white. The tail is black when new, all feathers tipped in white which abrades off during the breeding season. The bill and naked eye surround is bright yellow. The iris is dark brown over grey, with distinct white flecks. Legs are yellow brown and the claws horn-coloured. Juveniles are of paler plumage, the bill is light yellow streaked with dark grey, and the skin around the eye is white for the first two weeks. Juvenile tail feathers are without white tips and the iris is grey.

Voice: a loud *chickork-chickork-chickork* (territorial proclamation) given while head-bobbing, may be accompanied by a complicated series of quiet calls including *pee pee pee*, bell-like notes and deep guttural booming. Other quiet trills are used when inviting a mate to fly.

Similar species: there are no similar species in New Zealand; the common starling is the only other member of same family.

Distribution and Habitat

North of 40° S, mynas are found throughout the North Island and offshore islands but are absent from some areas of the volcanic plateau and large forested tracts such as Urewera National Park. They are also rare in the interior of forest remnants. Their natural distribution includes the entire South Asian region. They are now found on all continents with the exception of South America and Antarctica, and also many islands in the pacific, Indian, and Atlantic Oceans. They are widely regarded as serious invasive pests.

Population

The myna evolved to live in the tropics. It appears that the ambient temperature south of 40° S (i.e. Marton to Waipukurau) was insufficient to allow long term population survival. Some groups of mynas held on for many years around piggeries, but when whole milk collection replaced butter fat production and piggeries became rarer, the mynas could not balance their energy budget and died out. North of 40° S the population spread and increased, so that now in the north of the North Island mynas are one of our most conspicuous birds.

Breeding

Mynas nest in cavities, inside the eaves of houses, in letter boxes and cardboard cartons (even on the ground) and nesting boxes. Nest are made of dried grass, straw, cellophane and plastic and lined with green leaves - often eucalyptus leaves - just prior to laying. Nest preparation begins in late August to early September. Nests can be built in a week but usually take a few weeks. Laying begins in November and is finished by early February; Two peaks to egg laying representing first and second clutches occur in November and January. The few clutches between these two peaks are either replacement clutches, or clutches laid by inexperienced pairs. Clutch size averages 4 (1-6), the incubation period is 14 days; only the female has a true brood patch. Nestlings fledge 25 (20-32) days after hatching. Both parents feed the fledglings for 2-3 weeks, and approximately 20% of these die before independence. Productivity is about 2 chicks per annum per pair to independence.

Behaviour and Ecology

Mynas are not obviously sexually dimorphic. They pair for life but quickly find a new mate on the death of their previous mate. Both members of a pair proclaim ownership of a nest and territory using the head bob and loud call display described above, and vigorously defend their nest site and feeding territory against all other mynas. They destroy eggs and nestlings of any other species within their territory (especially starlings). Territories measured in Hawke's Bay averaged about 1.5 ha (0.6-2.3). In winter, they fly to communal roost up to 2 km distant. In the morning, they forage on roads then return to their nest site.

Food

Mynas are omnivorous. They consume pasture and crop invertebrates, including some that are crop pests. Also nightshade fruit and other fruit, especially berries. They peck at the flesh of ripe karaka fruit rather than swallowing the whole fruit. Mynas are most often seen along roads, where they seek road-killed insects. They often frequent refuse dumps in winter, seeking food scraps, and flock onto paddocks being ploughed. Mynas also take nectar, and are sometimes seen with orange flax pollen on their foreheads.

Damage by the Common Myna

The species is listed by the IUCN (World Conservation Union) as one of 100 of the world's worst invasive alien species alongside common starlings, red foxes and cane toads. It is a moderate pest of agriculture causing damage to orchard fruits such as fig, apple, pear, strawberry, guava, mango and grape. It also damages standing cereal crops including maize, wheat and rice.

In several countries it is considered an environmental pest and is reported to eat eggs and young birds and mammals including endangered species. The common myna aggressively competes for nest hollows and food, adversely affecting the breeding success of other birds and hollow-nesting mammals. It has been observed attacking terns on islands and taking the eggs of other seabirds, possibly interfering with those birds' breeding and general behaviour. The common myna is also known to spread avian malaria to other birds.

The species is a major disperser of seeds from the pest plant lantana and is suspected of spreading other environmental weed seeds such as olive.

It has the potential to transmit diseases to humans and can carry blood-borne parasites like plasmodium that causes malaria, as well as other parasites such as mites, roundworm and threadworm. Mites from common myna nests built in urban dwellings can invade homes and cause dermatitis and allergies in susceptible people.

In Australia programs involving community groups and government authorities have been developed to reduce common myna populations or damage. These programs have often been carried out in response to public concerns about the negative effects of the common myna on native birds. A variety of lethal and non-lethal methods have been used including; shooting and scaring with firearms, baiting, trapping, chemical and sound repellents, habitat and nest box modification.

Birds in the Wild

The common myna is prohibited in South Australia, Tasmania and Western Australia. Movement of the common myna to other areas can occur via shipping and road transport and/or possibly intentional release, and as a result it has been found most recently in Adelaide, South Australia; Launceston and Devonport, Tasmania; and Perth, Western Australia.

Moles

The Eastern mole (Scalopus aquaticus) is the first mammal found to smell in stereo.

Moles are small mammals that are found all over the world. They are often thought of as garden pests, mainly because of their intricate tunnel systems. And though they spend most of the time underground, they are not blind.

Size

These rotund animals have a hairless, pointed snout, small eyes and no visible ears. On average, moles grow to 4.4 to 6.25 inches (11.3 to 15.9 centimeters) long from snout to rump. Their tails add 1 to 1.6 inches (2.5 to 4 cm) of length. They typically weigh 2.5 to 4.5 ounces (72 to 128 grams).

The American species is a little on the larger side. The North American mole species tends to get as big as 7 inches (17.6 cm) long, 1.25 inches (3.3 cm) tall and weighs around 4 ounces (115 grams).

Habitat

Moles are found on every continent except Antarctica and South America. They live in grasslands, urban areas, gardens, grasslands, sand dunes, mixed woodland or any area that has soil where they can dig tunnels. They do tend to stay away from areas with acidic soil and mountainous areas, though.

Moles use tunnels to travel, but tunnels are more than just underground highways. Moles dig special chambers at the ends of tunnels that serve as bedrooms and birthing areas. Sometimes moles will live in a series of tunnels for generations before moving.

Moles have kitchens, as well, in tunnel chambers. They eat mostly earthworms, and keep them alive and immobile by biting their heads, and then store them in the chamber. As many as 470 worms have been recorded in one chamber.

Habits

Moles spend most of their lives alone and underground in their tunnels. Moles are such loners, in fact, that three to five moles per acre (7 to 12 hectares) is considered a lot.

Moles spend their time digging tunnels and hunting for food. A permanent tunnel is usually about 2 inches (5 cm) in diameter and 8 to 12 inches (20 to 30 cm) below the surface, while temporary tunnels are usually right under the surface of the ground.

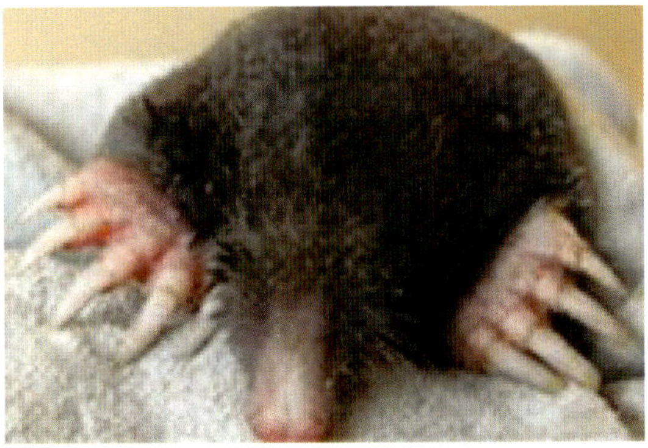

European mole (Talpa europaea)

Diet

It is a misconception that moles burrow into gardens to eat the roots of plants. They are actually after the earthworms that are found in garden soil. Moles love earthworms so much that they eat nearly their body weight worth of earthworms per day. For example, a mole weighing 2.8 ounces (80 g) eats around 1.7 ounces (50 g) of earthworms per day, according to the Mammal Society. Moles also consume insect larvae.

Offspring

During breeding season, males will enlarge their tunnel to more territories to find females to mate with. Once the breeding is done, a spherical nest chamber lined with dry plant material is created.

A female mole gives birth to three to four hairless babies at a time. By 14 days old, the mole babies, called pups, will start to grow hair. At four to five weeks, the pups are weaned, and at 33 days they leave the nest. By five to six weeks, pups leave their mother and their home tunnel completely. Moles typically live three years, according to YPTE.

Mole Removal and Control

Moles are usually classified as a pest species due to their tunneling habit. The most common complaints include the following:

- "Ridges" from tunnels all over yard
- Large mole hill piles of dirt
- Ground feels spongey & looks unsightly
- General lawn & landscape destruction

For these reasons, many people wish to have this nuisance animal removed from the property. This is usually accomplished with lethal traps - there are no known effective live mole traps.

Mole Biology: The Eastern Mole (Scalopus aquaticus) is small, weighing only about 3 oz. and is about 6 inches long. They live for about 2-3 years. They breed in January, and after a 45 day pregnancy, give birth to 2-4 young. The Townsend's Mole (Scapanus townsendii) is larger, at about 5 oz. and 8 inches long. Breeding is similar to other moles. The Star-Nosed Mole (Condylura cristata)is yet another potentially destructive mole species.

Mole Behavior: All moles are subterranean animals - they live under ground - and dig a network of tunnels and chambers. They create surface tunnels and dirt mounds under the chamber areas. Most moles are territorial, so there might just be one in your yard. They primarily eat earthworms, and also other underground larvae and insects.

Nuisance Concerns: It all has to do with the digging. Many people simply don't like the destruction to their lawn and landscaping. A single mole can turn a nice manicured lawn into a mess of tunnels and dirt piles. They wreck your yard.

Mole Diseases: No important zoonotic diseases associated with moles.

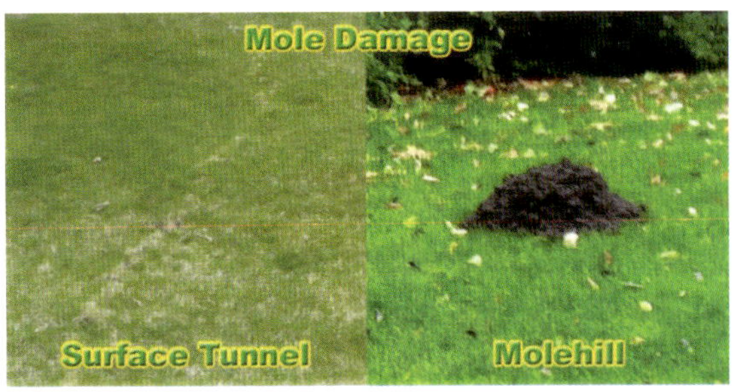

Methods to Get Rid of Moles

Lethal trapping and removal is the one and only way. Types of mole traps include spear traps, scissor traps, and body clamp traps. These traps are not for amateurs to set. Mole trapping is a subtle art. The traps have to be set just right or else they won't work. They have to be set on the active tunnels that the mole is using, not the one-time feeder tunnels. Mole trapping requires experience. I have to point out that there is no such thing as a live catch mole trap. These animals dig underground, and no one has yet developed a trap that can effectively catch them alive underground. Thus, all mole traps currently sold are lethal traps that kill moles. There are three major types of mole traps:

Spear Traps - These are set on top of an active surface tunnel. Squish the tunnel down, and set the spear trap on top, with a trigger pan against the dirt. When the mole crawls through the tunnel, it will push open the squished down area, and this will push up the trigger pan, causing the spears to slam downward, impaling the mole and killing it.

Scissor Traps - These traps are set over an active surface tunnel, with the scissor blades in the ground, on each side of the tunnel. When the mole crawls through the tunnel again, it triggers the trap, and the blades snap together, killing the mole.

Paper Clip Traps - These body clamp traps are actually set down inside the tunnel, often back to back, each facing outward. When the mole crawls through the tunnel as usual, coming from either direction, it will hit a pin holding the paper clip trap open, causing the pin to slip and the trap to slam shut on the mole, killing it.

Location - Finding the right mole tunnels to set the traps on is imperative. The more active, the better. Straight tunnels, or ones against edges (like a house or sidewalk) are a good bet for repeated use. The branching, winding tunnels that stretch out to nowhere might just be one-time use feeding tunnels, and thus not heavily travelled and not as productive for trapping.

Pigeons

Pigeons, also known as Rock Doves, are monogamous and will lay one to two eggs which hatch within eighteen days. Young are fed “pigeon milk” which is regurgitated food from both males and females. While mating occurs year round, the peak mating seasons are spring and fall.

Habits

Pigeons prefer grains for food and people will commonly feed pigeons unintentionally by spilling food or it in open trash containers. Pigeons roost in areas above ground and will readily nest in steeples, as well as voids on outside areas of buildings and other protected areas.

Habitat

Pigeons are dependent on humans to provide them with food, roosting and nesting sites. They are commonly found around agricultural areas as well as warehouses, feed mills, and grain elevators. They are also commonly found in cities around parks, buildings, bridges, and any other structures.

Threats

Pigeons are filthy birds, causing disease and damage. Their droppings are known for triggering human slips and falls, as well as accelerating the aging of structures and statues. This makes it imperative to get rid of pigeons in highly trafficked areas. More seriously, pigeons may carry diseases such as cryptococcosis, toxoplasmosis, salmonella, food poisoning, and more. Also, their droppings may harbor the growth of fungus, which causes histoplasmosis. Other pests may live on these birds, including fleas, lice, mites, ticks, and other pests. Pests may also infest nests such as stored product pests.

Pigeon Control Methods

There are many techniques for excluding pigeons from an infested structure. Exclusion strategies include structural modification, bird netting, plastic and metal spikes, monofilament and steel lines, and trapping. In addition, gels, paste, and liquids can also be used for excluding the birdsfrom the structure.

Structural Modification: Birds like to land on flat surfaces. When the building ledges are angled (45 degrees), it prevents birds from landing on building edges. Sheet metals, Styrofoam blocks, wood, stone, and other materials can be used to give the desired angle.

Bird Netting: There are several types of bird nettings that can be used for exclusion. In general, netting provides the best long-term control in keeping the birds out of a structure. In warehouses and commercial buildings where bird activity is a major concern, bird nets can be used to prevent the birds from getting into overhead pipes, roof air conditioning units can be netted off from the birds, etc.

Plastic and Metal Spikes: Spikes can be used as a physical barrier in preventing the birds from landing on an area, such as perching on ledges and beams of buildings. Spikes have sharp pointed edges extending out at angles. They can be installed around ledges, roof peaks, eaves, window sills, or any places that are prone to roosting. Because netting and spikes are more effective long-term control methods, they are the preferred methods.

Pigeon Repellent

Pigeon repellents are available in different formulations such as gel, paste and liquids. They differ considerably in cost, effectiveness, and texture (thick and tacky, jelly like or viscous).

Choose the one that works best based on the experience. These repellents are not non-toxic and because of their "sticky" properties cause the birds to move elsewhere.

Rabbits

Rabbits thrive in dryland and semi-arid environments, especially in New Zealand's South Island. They can populate these environments in truly destructive numbers if uncontrolled. Reduction of populations is on-going and essential for the economic and environmental welfare of New Zealand.

Rabbit

Facts about Rabbits

Domestic rabbits were commonly carried on sailing ships and sometimes released on islands as food for castaways. Rabbits were still imported as late as the 1860s to stimulate a domestic fur and meat trade. Establishment of wild populations was limited at first by lack of suitable habitat but increased pastoral farming assisted spread and growth in numbers. By the 1880s rabbits had become a serious threat to the fragile New Zealand economy.

Some wild rabbits may live up to seven years but life spans are generally much shorter, with high rates of natural mortality among young animals.

Rabbits have a high capacity for reproduction and female rabbits (does) may be pregnant for 70% of a year. Early-born does may breed in their natal year. They can produce a total of 20 – 50 young per adult doe. Females are also capable of adjusting litter-sizes to food supply so rabbit populations are capable of rebounding quickly from natural disasters or control pressures.

The Rabbit Threat

Rabbits are regarded primarily as an agricultural pest. They compete very effectively with livestock for pasture. On average, 7-10 rabbits consume as much as one ewe. Rabbits also provide a stable food source for mammalian carriers of bovine tuberculosis. Burrowing and scrapes cause extensive damage on erosion-prone soils, so much so that agricultural land can be rendered useless. Historically, rabbit-infested farms were abandoned because owners could not make a living.

Rabbits also threaten ecological values where they browse on vulnerable native plant communities, and where as year-round prey they support the mammalian predators which are contributing to the extinction of many New Zealand native birds and animals.

Devastation from rabbit burrows

Rabbit Damage

Rabbits are voracious eaters and leave clean-cut damage. Check the leaves and stems of your plants for cleanly cut damage; insects and other pests usually leave jagged edges on damaged plants. This clean-cut damage often happens at ground level, as rabbits tend to eat the yummy green shoots of tulips and other plants.

These low mowers graze close to the ground and sniff out the first tender young shoots and crop them short. They love to munch on flowers, clover, peas, lettuce, beans, and more. Many of these plants are also the favorites of woodchucks or groundhogs, so check for burrows before deciding you have rabbit damage. Once your plants have passed the seedling stage, they are usually safe from rabbit damage.

Although bunny nibbling occurs in every season, it's especially discouraging in the early spring when rabbits mercilessly munch the tender green shoots of plants.

Control and Prevention

How to get Rid of Rabbits

- As their twitching noses indicate, rabbits sniff a lot. Try sprinkling dried sulfur around or on your plants. Rabbits also dislike the smell of onions, so try planting these around your garden to further deter the furry creatures.
- To discourage pesky rabbits, try dusting your plants with plain talcum powder.
- Since rabbits are great sniffers, powdered red pepper sprinkled around the garden or on targeted plants may keep them out.
- Irish Spring soap shavings placed in little drawstring bags around the garden will also help to keep rabbits away.
- Make a bad-tasting rabbit cocktail by grinding together three hot peppers, three large onions, and one whole bunch of garlic. Add water to cover, and place into a covered container overnight. Strain, and then add enough additional water to make a gallon of the mixture. Spray onto plants, repeating after rainfall. Commercial products using pungent garlic oil are also worth a try.
- Spray your plants with a mixture of 1 teaspoon Lysol and 1 gallon of water.
- Some people protect plants with individual "collars" of tin cans or screening so that the plants may reach a less vulnerable size. Put the collar around each stem for protection.

- Use cylinders of 1/4-inch mesh hardware cloth to keep rabbits from nibbling on young fruit and landscape trees. The cylinders should extend higher than a rabbit's reach while the rabbit is standing on the expected depth of snow, and they should stand one to two inches out from the tree trunk.
- Some of the deer techniques related to odor are also said to work against rabbits. Deter rabbits with commercially-available deer repellents that contain a mixture of dried bovine blood, sulfured eggs, and garlic.
- Legend has it that rabbits are terrified of their own reflection, so try an old-time rabbit remedy and place large, clear glass jars of water throughout the garden. Garden centers sell ready-made reflectors, as well as other devices—crouching cats, fake snakes, menacing owls—designed to frighten bunnies away from your plants.
- Sometimes, humane traps are the best solution. If you don't want to buy a trap, consider building one. Place the trap where you've seen the rabbits feeding or resting, and cover it with a piece of canvas. Apples, carrots, cabbage, and other fresh green veggies make excellent bait. Check it often, and release bunnies in rural areas several miles away.

Keeping Rabbits Away

The best way to keep rabbits out of the garden is to start early in the spring using the things they don't like, then be consistent throughout the growing season.

- It's best to keep rabbits from crossing into the garden to begin with, and many old-time remedies rely on spreading various products around the perimeter of the garden such as dried blood or dried blood meal or human hair. Sprinkle dried blood on the surface around all your plants as early in the season as you can, and repeat after a heavy rain.
- Do not rely on repellents alone. The most effective way of keeping out rabbits is chicken wire fencing. Install a fence that is 4 feet high and bury it 6 inches deep. Bend the top foot of the fence away from the garden like a security fence, so that they can't climb or jump over it. For new bulbs, try a dome or cage of chicken wire secured over the bed.

- Rabbits don't like to leave their shelters, so try to reduce the possible rabbit homes around your yard. Brush away piles of brush and leaves, and fill in abandoned burrows. If a rabbit doesn't have a place to live, hopefully it won't stay and munch. Rabbits will also breed much more if they have a good habitat available—all the more reason to have no vacancy.

Plants that Rabbits Dislike

According to bunny experts, rabbits have plant preferences based on taste, nutritive value, the presence of poison or prickles, and ease of availability. Their tastes in food can also vary by region and season, so not all plants work for all rabbits. Be tricky and tend plants that rabbits don't find very appetizing.

Rabbits tend to avoid some of the same plants as deer and Japanese beetles. If you'd like to control all these pests, check our list of deer-resistant plants and best and worst plants for Japanese beetles to know which plants might do best. Choose plants such as forsythia, lilac bush, marigolds, zinnias, daffodils, lavender, and snapdragons for rabbits. This might help to reduce your rabbit population. This is not a guaranteed solution, as hungry rabbits will eat almost anything, but filling your garden with these plants might make your garden less appetizing than another one.

References

- Shōzō Ōsawa, Zhi-Hui Su & Yūki Inmura (2004). Molecular Phylogeny and Evolution of Carabid Ground Beetles. Springer. ISBN 4-431-00487-4
- Gomes Gonçalves, Marcos Paulo (December 2017). "Relationship Between Meteorological Conditions and Beetles in Mata de Cocal". Revista Brasileira de Meteorologia. pp. 543–554. doi:10.1590/0102-7786324003. ISSN 0102-7786
- "Trees of family Carabidae". Carabidae of the World Database. 2008. Archived from the original on September 19, 2008. Retrieved July 24, 2008
- Bura, V. L.; Rohwer, V. G.; Martin, P. R.; Yack, J. E. (2010). "Whistling in caterpillars (Amorpha juglandis, Bombycoidea): Sound-producing mechanism and function". Journal of Experimental Biology. 214 (Pt 1): 30–37. doi:10.1242/jeb.046805. PMID 21147966
- Donna Spalding Andréolle & Veronique Molinari, eds. (2011). Women and Science, 17th Century to Present: Pioneers, Activists and Protagonists. Cambridge Scholars Publishing. p. 36. ISBN 9781443830676
- Redd, JT; Voorhees, RE; Török, TJ (2007). "Outbreak of lepidopterism at a Boy Scout camp". Journal of the American Academy of Dermatology. 56 (6): 952–955. doi:10.1016/j.jaad.2006.06.002. PMID 17368636
- Pierce, N.E. (1995). "Predatory and parasitic Lepidoptera: Carnivores living on plants" (PDF). Journal of the Lepidopterist's Society. 49 (4): 412–453
- Hall, Donald W. (September 2014). "Featured Creatures: hickory horned devil, Citheronia regalis". University of Florida, Entomology and Nematology Department. Retrieved 19 September2017

- Himmelman, John (2011). Cricket Radio. Harvard University Press. p. 45. ISBN 978-0-674-06102-6. Archived from the original on 2017-11-27
- Heitler, W. J. (January 2007). "Energy and Power". University of St Andrews. Archived from the original on 18 November 2014. Retrieved 5 May 2015
- Umbers, K.; Tatarnic, N.; Holwell, G.; Herberstein, M. (2012). "Ferocious Fighting between Male Grasshoppers". PLOS ONE. 7(11): e49600. doi:10.1371/journal.pone.0049600. PMC 3498212. PMID 23166725
- Bennet-Clark, H. C. (1975). "The energetics of the jump of the locust Schistocerca gregaria". The Journal of Experimental Biology. 63 (1): 53–83. PMID 1159370. Archived from the original on 2017-01-03

Chapter 3

Bird Pest Control

The methods that are used to deter or eliminate pest birds from nesting, roosting and landing fall under bird control. This chapter closely examines the varied ways to ensure bird control, such as by using bird control spike, bird netting, avicide, bird scarer and scarecrow.

Bird Control

Birds are naturally drawn to food supplies in agricultural areas, transmitting diseases and damaging crops. Fruits, vegetables and newly-seeded and emergent crops provide an all-you-can-eat feast for birds if unchecked. Birds typically eat their body weight in food every day causing damage to crops and food storage in barns ranging from 5-50% or more.

Globally, this problem causes billions of euros of damage each year. Effective bird control is essential and farmers do anything to keep birds away from their crops.

Following are the typical bird problems one encounters:

- Bird fouling – deposits on buildings, solar panelling cars, stock, rooftop plant machinery or walkways can be a significant hazard and health and safety risk
- Aggressive behaviour – very common during the nesting season (March – September)
- Nesting in or on buildings – clogging drains, flues encouraging insect infestation
- Noise – Some bird species are very loud, for example the gull calls can be disruptive to business or residential areas
- Ground nesting – ground nesting bird species can cause significant development delays and costs if not managed correctly or legally

Other problem bird species that come into conflict with businesses and organisations are canada geese, starlings and corvids (crows, rooks, ravens, jackdaws) which create health and safety issues, depositing bird mess and creating destruction wherever they go.

Use an Integrated Approach

The most effective bird control uses a combination of deterrent methods at the same time. Even netting does not provide 100% protection and can be improved with other devices. There are three categories of bird deterrent methods currently available:

- acoustic
- visual
- physical

Chemical deterrents (e.g., methyl anthranilate) are also available. These products must be approved for use on the specific crop and used according to the label instructions. These products can leave unwanted flavours on fresh food. As a result, their use in agricultural applications is very limited.

Acoustic Deterrents

Acoustic deterrents use sound to scare birds away. Birds have a hearing range similar to humans; if people can hear it, so can birds. Birds cannot hear ultrasonic sounds.

Propane-fired Cannons (Bird Bangers)

Bird bangers ignite propane to produce loud, random, unexpected sound blasts (figure below). Modern devices fire sets of 3 blasts while rotating to cover a larger area. They must be used in compliance with best management practices.

A propane-fired canon.

Electronic Sound Devices (Squawkers)

Electronic sound devices broadcast either electronic noises to irritate birds and disrupt their sensory system, or distress calls of specific bird species and predatory attack calls (figure below). The latter sound like real birds and are usually more acceptable to neighbours than propane-fired cannons. Some farmers report the distress calls actually attract birds of prey, which can help scare problem birds away.

Electronic sound devices and propane-fired cannons are often used together. The electronic devices keep birds irritated and edgy, while the cannon provides stimulus to drive them away.

Pyrotechnic Pistol Cartridges

Pyrotechnic cartridges are launched from a handheld pistol and "fly" directly into a flock of birds, where they explode and produce a loud sound and/or visual effects. They can quickly scare away an entire flock. However, they must be manually operated. Take care when using cartridges near neighbours and observe the same setbacks (from where the cartridges explode) and guidelines as propane-fired cannons.

Shotguns

Shotguns pose a serious safety risk and must never be used without a licence and proper training. A shotgun is not as effective at scaring birds as a pyrotechnic cartridge. Lethal control is usually ineffective because of the large bird numbers. Obtain permission from governing authorities before using shotguns.

Other Sound-Producing Devices

Devices such as air horns, clanging metal or Mylar humming lines generally work for a few days before birds start to ignore them. They can also be disruptive to neighbours. These devices may provide the most benefit when used only in the days just before harvest, when bird pressure is highest.

An electronic sound device (squawker)

Best Practices for using Acoustic Bird Deterrents

- Inform neighbours
    - Explain which devices are being used, how they work and when they will operate.
    - Explain how to reach you if there is a problem.

- Use sparingly during nesting
  - Use electronic bird scarers and visual deterrents in nesting areas.
  - Stop after nesting season.
- Wait for ripening fruit
  - Do not use deterrents until 3-4 weeks before the crop starts to ripen (depends on crop).
- Dawn to dusk only
  - Use only between sunrise and sunset when birds feed.
  - Use electronic timers and light sensor overrides.
- Observe minimum setbacks
  - at least 125 m from neighbours' houses
  - at least 125 m from other acoustic devices
- Use proper frequency
  - Use the 4-8-minute setting on propane-fired cannons.
- Move devices regularly
  - Move devices at least once a week.
  - Move devices immediately if there is no effect.
- Stop as soon as possible
  - Stop as soon as crops are harvested.
  - Stop or move devices immediately if there is no effect.

Bird Control Spike

Methods of deterring birds, like spikes and netting, have a sole purpose to keep the birds away without harming. Not only from a legal and environmental perspective, is killing birds in large numbers also inhumane and necessary. Bird spikes are often used as a solution in keeping birds away from spaces that are normally hard to reach, or too large to net. They are amongst the humane deterring methods, since they don't cause the animal any pain or distress. Compared to other deterrents, like shiny objects, artificial animals, electronic boxes and shock tracks, bird spikes seem to be a more humane solution. Bird spikes will simply not allow a bird to land on to the area, without scaring it or causing it any distress. When it comes to this particular method of deterring birds, there are several important facts to know before you proceed to install them. In order for these to be effective, they not only need to be installed properly, but they also need to be used for the adequate species of birds.

Bird spikes are a tool that consists of long, pointed spikes that are lodged into the basis. They are normally made out of steel, metal or plastic. Bird spikes are most often installed in areas that are either too large to cover with bird nets, or too inaccessible to use other deterrents in. open spaces, like tall and long fences surrounding commercial and industrial buildings, are normally hard to protect against birds using any other method. Fogging in this case simply doesn't make sense, and setting up nets and other tools like artificial animals simply might not be appropriate for these types of spaces. Additionally, spaces like balconies, soffits and many other that are hard to cover with other deterring tools, are much easier to bird-proof by installing bird spikes. Once the spikes are set up in these areas, birds will no longer be able to land, because the spikes are too thin and sharp. Eventually, birds will lose interest in these areas.

Bird spikes serve to deter almost all of the bird species, including pigeons, sparrows, crows, starlings, seagulls, and many other. This means that birds spikes are appropriate to use in both open and closed areas, as well as areas where one might themselves in a problem with marine bird roosting. In addition, bird spikes are quite simple and easy to install, and they are also very affordable, long-term investment in the safety and comfort of an estate. These, however, don't come without serious shortcomings. The biggest amongst the problems include piling up of bird feces in the spikes. While hovering over the deterrent and trying to find a place to land, birds will drop substantial amounts of feces. It is possible that the feces completely fills up areas of the spikes, to the point where birds will be able to land again. This means that you are either going to have to change them from time to time or clean them regularly.

Bird spike installation is a quite simple process. First, you need to purchase an adequate amount of spikes from a supplier. To know just about how much of these you will need, you need to know the complete length of the area you wish to cover. After you decide on the amount of spikes you need, you can choose between different types of spikes that vary in size and material. Spikes are usually made out of either plastic, steel or metal. The difference is usually in the durability of the material, meaning that you can expect the steel ones to last you the longest. The installation process consists out of drilling the basis of the spikes and screwing them into the surface. After this, the only thing you need is to make sure that the screws are sturdy and that spikes are not moving around. Depending on the material from which they are made of, bird spikes can last for longer or shorter periods of time.

Bird Netting

Bird netting excludes nuisance birds from areas such as overhangs, eaves, warehouses, rooftops, and airplane hangars and enclosed areas where pest birds need to be kept out. Bird netting provides 100% exclusion of the nuisance birds and is a method that can be used long-term. The netting comes in many different mesh sizes which is ideal for different types of birds. For example, 2" mesh is used for larger pest birds including pigeons, seagulls and crows. There are also different colors of bird net including black, white and stone color. It will match the environment it is used in to help its invisibility factor that it already has when it is installed correctly. Many choose the black bird net as it is naturally U.V. protected. It is also rot-proof, waterproof and flame resistant.

There are different types of netting that can be purchased. For example, there is heavy duty bird netting that is a knotted polyethylene. This type of netting has a high melting point and is flame resistant. It is ideal in large areas where pest birds are a problem such as warehouse, canopies, airplane hangars and other overhangs. This type of netting is non-conductive, meaning it can be installed around electrical systems and RF transmissions systems without any interference. The netting is made up of six monofilaments with each strand 0.0012 thick and twisted together to create a strong twine with 160-200 twists per meter.

The different types of anti-bird netting come in different mesh sizes. It is important to choose the correct mesh size for your particular bird issue. It is critical that birds are unable to get past the netting and get trapped behind the netting or stuck in the netting. The larger the bird, the larger the mesh size used.

Garden netting is another type of anti-bird net used in the garden and around the home. This type of bird control netting is also used in vineyards. It has a light-weight plastic mesh. It is environmentally friendly, made from FDA sanctioned resin. It is UV stabilized to endure months in the sun or rain. When it is installed correctly it will blend into the bushes, plants, vines and trees.

Quality anti-bird netting is a must. Reputable companies will also have qualified experts and authorized installers available to help with your every need.

Avicide

When people contact pest control companies for advice on how to resolve conflicts with birds, operators often recommend using avicides, or bird poisons, such as Avitrol, DRC-1339, and others to "control" bird populations.

Companies often tell potential customers that Avitrol and other avicides are "flock-dispersing agents" that "scare" birds away from areas where they are not wanted. They'll even tell customers that these products are humane. Nothing could be further from the truth.

Avicides like Avitrol are acutely toxic and cause birds and other animals to suffer immensely. Avitrol attacks and impairs birds' nervous systems, causing birds who ingest the poison to become disoriented, exhibit erratic flight and tremors, and suffer violent convulsions for hours before they finally succumb to the effects of the toxin.

Bird Scarer

Bird scarers have been used in agriculture and horticulture for many years as a method of reducing damage to crops.

Two main types are in use:

- Explosive – which use propane gas or a cartridge to cause a loud explosion
- Amplified sirens or recorded distress calls

Both types make a noise which is likely to be heard over a wide area, with the potential to cause disturbance to local residents. The explosive type give rise to the majority of complaints.

Effectiveness of Scarers

All bird scarers can become familiar to birds over time and lose their effectiveness. To help avoid this, they need to be regularly moved and used with other methods to scare birds and protect crops.

- Use non-audible scarers wherever possible

- Use other protection techniques where practicable e.g. netting, patrolling
- Use acoustic scarers, only when there is an actual risk of damage to crops
- Pay careful attention to siting of scarers – allow at least 200 metres from the nearest house
- Arrange to direct the sound from the scarers into the area to be protected, and away from houses, by using simple baffles such as straw bales
- Limit the number of scarers at a single location – allow at least 100 metres between scarers
- Control the frequency of explosions to a maximum of four firings per hour. Multiple firings from a multiple chamber gun count as one firing if heard within 30 seconds
- Do not use auditory scarers between sunset and sunrise
- Do not mount explosive scarers at high level

Visual Scarers

Scarecrow

One of the oldest designs of bird scarer is the scarecrow in the shape of a human figure. The scarecrow idea has been built upon numerous times, and not all visual scare devices are shaped like humans. The "Flashman Birdscarer," Iridescent tape, "TerrorEyes" balloons, and other visual deterrents are all built on the idea of visually scaring birds. This method doesn't work so well with all species, considering that some species frequently perch on scarecrows.

Hawk Kite

A stationary modelled owl used as a bird scarer at Lysaker, Norway.

Many species of bird are also naturally afraid of predators such as birds of prey. "Hawk kites" are designed to fly from poles in the wind and hover above the field to be protected. They are shaped to match the silhouette of a bird of prey.

Helikites

The Helikite bird scarer is a lighter-than-air combination of a helium balloon and a kite. Helikites fly up to 200vft in the air with or without wind. Although they do not look like hawks, they fly and hover high in the sky behaving like birds of prey. Helikites successfully exploit bird pests' instinctive fear of hawks and can reliably protect large areas of farmland.

Lasers

The use of lasers can be an effective method of bird scaring, although there is some evidence to suggest some birds are "laser-resistant". As the effectiveness of the laser decreases with increasing light levels, it is likely to be most effective at dawn and dusk. Although some lasers prove to be effective during daylight hours.

The method relies on birds being startled by the strong contrast between the ambient light and the laser beam. During low light conditions this technique is very selective and can be attuned to frequencies and wavelengths that individual bird species don't like, but at night the light beam is visible over a large distance and can cause widespread (non-species specific) disturbance. Lasers use can be limited due safety concerns of the beam and some nations have laws which prohibit lasers above a certain power from being used. Lasers should be looked at as an additional scaring system to add to a wildlife management program.

Manually operated laser torches and automated laser bird deterrent robots that move the laser automatically towards the birds are available on the market.

Dead Birds

The use of model or actual dead birds is used to signal danger to others. Initially, birds often approach the corpse but usually leave when they see the unnatural position of the bird. This approach has been frequently used in attempts to deter gulls from airports. Pheasant feed sacks often have an image of an owl with large eyes so that when empty they can be strung up to scare predators.

Balloons

Balloons are an inexpensive deterrent. However, this method relies on the movement of balloons, which is something that birds can become used to. The addition of eye illustrations on the balloons has been shown to increase this method's effectiveness as it combats the birds' ability to adapt. Commercially available "scare-eye" balloons have holographic eyes that follow birds wherever they go. The long-term effectiveness of this method can be increased by periodically moving the placement of the scare devices.

An example of a visual bird-scare balloon

In the United Kingdom the use of balloons is subject to approval from the Civil Aviation Authority, especially around airfields.

Auditory Scarers

Audible bird scarers use noise stimuli that makes birds uncomfortable. However, once birds realize these pose no real threats, they can easily become habituated to sounds that seemed initially frightening. If just being placed in situ and left, audible bird scarers can easily become ineffective bird control solutions, however when managed on an ongoing basis or used as part of a greater bird deterrent system, sound methods can deliver quality results.

Sōzu (Shishi Odoshi)

One very old design is the Japanese sōzu, known metonymically as a shishi odoshi (although the term *shishi odoshi* properly refers to any method of scaring wild animals, including the Western scarecrow). Instead of using a visual method to distract pests, as the scarecrow does, it uses the sound of a heavy pipe repeatedly and rhythmically hitting a rock, using water as a timing device. The sōzu is also used in Japanese popular culture to denote inordinate amounts of wealth, combined with a traditional sensibility: by design, the shishi odoshi uses copious amounts of water, meaning either a very high water bill, or that it is situated on high-value land with a stream or river running through it.

Propane Cannons

Propane scare cannons are one of the most common types of bird scarer available in Europe and America. The propane powered gas gun which produces a periodic explosion. The audible bang can reach very loud volumes in excess of 150 decibels, causing a flight reaction in birds. The similarity between a scare cannon and a 12 gauge shotgun is thought to cause a startle/fear reaction, although it is also effective against birds that have not been exposed to hunting pressure.

A typical propane gas gun bird scarer.

Birds can become habituated to the sound of regular cannon detonations, especially if it does not vary in its magnitude, pitch, or time interval. However, regularly moving the cannon, utilizing on-demand firing options, including radio control, and combining cannons with other methods of deterrents can prevent habituation.

Propane Scare Cannons are very loud and can be disruptive to people living nearby. One study found that by restricting cannon use to only hours when birds are active and incorporating better bird damage plans drastically reduced the number of complaints from neighbours.

Electronic Repellers

Sonic bird repellers are not effective; the birds quickly acclimate to them. Usually consisting of a central unit and several speakers, the system emits digitally recorded distress calls of birds, and, in some cases, calls of predators of the target species. Some emitters randomize pitch, magnitude, time interval, sound sequence and other factors in an attempt to prevent birds from getting used to them. Many of the sounds produced are regarded as annoying to people.

Ultrasonic Scarers

Ultrasonic devices are static sound-emitting bird deterrents, which, in theory, will annoy birds to keep them away from enclosed or semi-enclosed areas. Ultrasonic scarers are not harmful to birds, however, there is debate around birds' ability to hear these frequencies at loud enough decibels. Birds are believed to have similar hearing to humans, with studies showing birds do not hear on an ultrasonic level,meaning that ultrasonic scarers often have little or no effect in deterring birds.

Cartridge Scarers

Cartridge scarers include a wide variety of noise-producing cartridges usually fired from rockets or rope bangers, or on aerodromes from modified pistols or shotguns, which produce a loud bang and emit flashes of light. They include shellcrackers, screamer shells and whistling projectiles, exploding projectiles, bird bangers and flares. Bird banger cartridges commonly use a low explosive known as flash powder.

Cartridges are projected from a shotgun with a range of 45–90 metres (148–295 ft), or pistols with a range of approximately 25 metres (82 ft), before exploding. Bird scaring cartridges can produce noise levels of up to 160 dB at varying ranges but in some countries both the cartridges and the gun require a firearms certificate.

Pyrotechnics have proved effective in dispersing birds at airports, landfill sites, agricultural crops and aquaculture facilities. At airports in the United Kingdom, shellcrackers fired from a modified pistol are the most common means of dispersing birds, as they allow the bird controller to have some directional control over birds in flight, so they can be steered away from runways.

However, as with all similar noises, there is a high risk of birds becoming used to any pyrotechnics or cartridge explosions.

Benign Acoustic Deterrence

In 2013, Dr. John Swaddle and Dr. Mark Hinders at the College of William and Mary created a new method of deterring birds using benign sounds projected by conventional and directional (parametric) speakers. The initial objectives of the technology were to displace problematic birds from airfields to reduce bird strike risks, minimize agricultural losses due to pest bird foraging, displace nuisance birds that cause extensive repair and chronic clean-up costs, and reduce bird mortality from flying into man-made structures. The sounds, referred to as a "Sonic Net," do not have to be loud and are a combination of wave forms - collectively called "colored noise" - forming non-constructive and constructive interference with how birds talk to each other.

Technically, the Sonic Nets technology is not a bird scarer, but discourages birds from flying into or spending time in the target area. The impact to the birds is similar to talking in a crowded room, and since they cannot understand each other they go somewhere else. Early tests at an aviary and initial field trials at a landfill and airfield indicate that the technology is effective and that birds do not habituate to the sound. The provisional and full patents were filed in 2013 and 2014 respectively, with further research and commercialization of the technology being ongoing.

Other

Dogs

The control of birds and other wildlife such as deer through harassment by trained border collies has been used at aerodromes, golf courses and agricultural land. The dogs represent an actual threat, and so elicit flight reactions. Habituation is unlikely as they can continually pursue and change their behaviour. Border collies are used as they are working dogs bred to herd animals and to avoid attack, and they respond well to whistle and verbal commands. A single border collie and its handler can keep an area of approximately 50 square kilometres (19.3 square miles, 4998.7 hectares, or 12,179.2 acres) free of larger birds and wildlife and although they are effective at deterring ground foraging birds such as waders and wildfowl, they are not so useful for species that spend most of their time flying or perching, such as raptors and swallows.

In 1999 Southwest Florida International Airport became the first commercial airport in the world to employ a border collie in an airfield wildlife control programme. After the use of the collie, numbers and species of birds on the airport declined and most birds that remained congregated in a drainage ditch away from the runway. The number of bird strikes dropped to zero compared to 13 for the same period the previous year. Several other airports and airbases have now started similar programmes.

At Dover Air Force Base, Delaware, bird strike damage to aircraft caused by birds has been reduced from an average of US $600,000 / year for the proceeding two years to US$24 000/year after the initiation of a bird control programme that included the use of border collies.

Predators

Harris's hawk

Using predators as a natural bird deterrent has become a recommended form of controlling bird infestations. Specially selected species are trained to deal with working in un-natural environments with distractions and dangers they would not usually encounter.

The success of this method of bird control is based on the fact that many birds have a natural fear of falcons and hawks as predators, so their presence in the area encourages problem species to disperse. The natural reaction of most prey species is to form a flock and attempt to fly above the falcon. If this fails, they will attempt to fly for cover and leave the area.

Radio Controlled Aircraft

Radio-controlled model aircraft have been used to scare or ‹haze' bird pests since the early 1980s, mainly over airfields, but have also been used over agricultural areas, fisheries and landfill sites. This method has been shown to be very effective and birds habituate more slowly to a treatment in which they are being actively hazed. At Whiteman Air Force Base, Missouri, balsa wood radio-controlled aircraft are one of the primary bird harassment methods used to keep the airfield clear of raptors and other large birds, and they have also proved effective at dispersing the base's redwing blackbird roost.

Fireworks

Fireworks can also be used as bird scarers, and some jurisdictions issue special licences for agricultural fireworks. This practice has been criticised as a loophole for the sale of consumer fireworks. Again, the loud bangs can also irritate people living on nearby properties.

Combination Scarers

These combine multiple deterrents, such as using a pop up scarecrow combined with a gas gun, which in turn activates the distress call of a bird. These combination scarers are often managed by computers and synchronised across an area via the use of radio links. This synchronisation becomes more effective if there is some kind of detection system involved such as bird detecting radar.

Scarecrow

When farmers plant seeds, they intend the seeds to grow. But seeds are food for a range of wild animals, including mice and birds. Flocks of birds often settle on land that has recently been ploughed. At this time it is easy to pick among the loosened soil to find the seeds. Once the seeds have grown into plants, farmers have to worry about the birds eating the seeds and fruits that the plants might produce.

As you will know, birds are wary animals, and will fly out of harm's way if something approaches them. They have learned that people can be dangerous, and so they fly off. For many centuries farmers have made use of this fact by putting straw figures made to look like people, and dressed in old clothes, out into fields. Here the idea is to frighten off the birds without people having to be present all of the time.

Scarecrows used to be common in fields for this reason, and many very interesting scarecrows have been made down the ages. But it is not so much the presence of a scarecrow that frightens off birds as the fact that people move and make unexpected sounds. So the scarecrow itself is not really the best way of keeping birds off. Modern scarecrows do not look like people at all. They are often discs of highly-reflective plastic that are put on poles so they catch the wind and spin. This sudden changes in light is meant to scare off birds. Many farms also use automatic noise guns powered by propane gas. The loud explosion will send a flock of birds into the air - for a while.

Scarecrow, device posted on cultivated ground to deter birds or other animals from eating or otherwise disturbing seeds, shoots, and fruit; its name derives from its use against the crow. The scarecrow of popular tradition is a mannequin stuffed with straw; free-hanging, often reflective parts movable by the wind are commonly attached to increase effectiveness. A scarecrow outfitted in clothes previously worn by a hunter who has fired on the flock is regarded by some as especially efficacious. A common variant is the effigy of a predator (*e.g.*, an owl or a snake).

Scarecrows in a paddy

The function of the scarecrow is sometimes filled by various audio devices, including recordings of the calls or sounds of predators or noisome insects. Recorded sounds of deerflies in flight, for example, are used to deter deer from young tree plantations. Automatically fired carbide cannons and other simulated gunfire are used to keep migrating geese out of cornfields.

Chapter 4

Pesticides and their Role in Pest Control and Management

Pesticides are chemical substances or biological agents used for the control of pests and weeds. Depending on the pests targeted, they can be classified as herbicides, molluscicides, acaricides, piscicides and dicofol. This chapter includes detailed information about such common pesticides and their role in pest control and management.

Pesticide

Pesticide is any toxic substance used to kill animals, fungi, or plants that cause economic damage to crop or ornamental plants or are hazardous to the health of domestic animals or humans. All pesticides interfere with normal metabolic processes in the pest organism and often are classified according to the type of organism they are intended to control.

The use of certain pesticides is controversial. The insecticide DDT, for example, which came into wide use in the 1940s, was later heavily restricted in the United States and elsewhere because of its adverse affects on the health of the environment, wildlife, and humans. In the early 21st century the use of neonicotinoids was highly restricted in some countries, including throughout the entire European Union, because of the possible involvement of those insecticides in the decline of honeybee populations.

Active Ingredients

The active ingredient, usually the only component of the formulation listed on the pesticide label, is by nature biologically and chemically active against a target pest, be it an insect, weed or fungus. By definition these chemicals kill living things.

Contaminants and Impurities

Contaminants and impurities are often a part of the pesticide product and responsible for product hazards. Dioxin and DDT have been identified as contaminants, which have not been purposefully added but are a function of the production process.

Metabolites

Metabolites are breakdown products that form when a pesticide is used in the environment and mixes with air, water, soil or living organisms. Often the metabolite is more hazardous than the parent pesticide.

Inert Ingredients

If you were to go to your local hardware store and take a look at the label on a can of ant and roach killer, the contents might read something like this, "5% Permethrin, 95% Inert Ingredients." After reading the label, you may wonder what makes up the other 95%. The fact is, the manufacturer doesn't have to tell you. Currently, under the Federal Insecticide, Fungicide and Rodenticide Act (FIFRA), pesticide manufacturers are only required to list the active ingredients in a pesticide, leaving consumers and applicators unaware of the possible toxics present in the inert ingredients of pesticide products they are using, unless the EPA administrator determines that the chemical poses a public health threat. Pesticide manufacturers argue they cannot release information on inert ingredients because they are trade secrets, and if released, their products could be duplicated. Quite often inert ingredients constitute over 95% of the pesticide product. Inert ingredients are mixed into pesticides products as a carrier or sticking agent, and are often as toxic as the active ingredient.

Alternatives

Alternatives to pesticides are available and include methods of cultivation, use of biological pest controls (such as pheromones and microbial pesticides), genetic engineering, and methods of interfering with insect breeding. Application of composted yard waste has also been used as a way of controlling pests. These methods are becoming increasingly popular and often are safer than traditional chemical pesticides. In addition, EPA is registering reduced-risk conventional pesticides in increasing numbers.

Cultivation practices include polyculture (growing multiple types of plants), crop rotation, planting crops in areas where the pests that damage them do not live, timing planting according to when pests will be least problematic, and use of trap crops that attract pests away from the real crop. Trap crops have successfully controlled pests in some commercial agricultural systems while reducing pesticide usage;however, in many other systems, trap crops can fail to reduce pest densities at a commercial scale, even when the trap crop works in controlled experiments. In the U.S., farmers have had success controlling insects by spraying with hot water at a cost that is about the same as pesticide spraying.

Release of other organisms that fight the pest is another example of an alternative to pesticide use. These organisms can include natural predators or parasites of the pests.

Biological pesticides based on entomopathogenic fungi, bacteria and viruses cause disease in the pest species can also be used.

Interfering with insects' reproduction can be accomplished by sterilizing males of the target species and releasing them, so that they mate with females but do not produce offspring. This technique was first used on the screwworm fly in 1958 and has since been used with the medfly, the tsetse fly, and the gypsy moth. However, this can be a costly, time consuming approach that only works on some types of insects.

Push Pull Strategy

The term "push-pull" was established in 1987 as an approach for integrated pest management (IPM). This strategy uses a mixture of behavior-modifying stimuli to manipulate the distribution and abundance of insects. "Push" means the insects are repelled or deterred away from whatever resource that is being protected. "Pull" means that certain stimuli (semiochemical stimuli, pheromones, food additives, visual stimuli, genetically altered plants, etc.) are used to attract pests to trap crops where they will be killed. There are numerous different components involved in order to implement a Push-Pull Strategy in IPM.

Many case studies testing the effectiveness of the push-pull approach have been done across the world. The most successful push-pull strategy was developed in Africa for subsistence farming. Another successful case study was performed on the control of *Helicoverpa* in cotton crops in Australia. In Europe, the Middle East, and the United States, push-pull strategies were successfully used in the controlling of *Sitona lineatus* in bean fields.

Some advantages of using the push-pull method are less use of chemical or biological materials and better protection against insect habituation to this control method. Some disadvantages of the push-pull strategy is that if there is a lack of appropriate knowledge of behavioral and chemical ecology of the host-pest interactions then this method becomes unreliable. Furthermore, because the push-pull method is not a very popular method of IPM operational and registration costs are higher.

Effectiveness

Some evidence shows that alternatives to pesticides can be equally effective as the use of chemicals. For example, Sweden has halved its use of pesticides with hardly any reduction in crops. In Indonesia, farmers have reduced pesticide use on rice fields by 65% and experienced a 15% crop increase. A study of Maize fields in northern Florida found that the application of composted yard waste with high carbon to nitrogen ratio to agricultural fields was highly effective at reducing the population of plant-parasitic nematodes and increasing crop yield, with yield increases ranging from 10% to 212%; the observed effects were long-term, often not appearing until the third season of the study.

However, pesticide resistance is increasing. In the 1940s, U.S. farmers lost only 7% of their crops to pests. Since the 1980s, loss has increased to 13%, even though more pesticides are being used. Between 500 and 1,000 insect and weed species have developed pesticide resistance since 1945.

Types

Pesticides are often referred to according to the type of pest they control. Pesticides can also be considered as either biodegradable pesticides, which will be broken down by microbes and other living beings into harmless compounds, or persistent pesticides, which may take months or years before they are broken down: it was the persistence of DDT, for example, which led to its accumulation in the food chain and its killing of birds of prey at the top of the food chain. Another way to think about pesticides is to consider those that are chemical pesticides are derived from a common source or production method.

Insecticides

Neonicotinoids are a class of neuro-active insecticides chemically similar to nicotine. Imidacloprid, of the neonicotanoid family, is the most widely used insecticide in the world. In the late 1990s neonicotinoids came under increasing scrutiny over their environmental impact and were linked in a range of studies to adverse ecological effects, including honey-bee colony collapse disorder (CCD) and loss of birds due to a reduction in insect populations. In 2013, the European Union and a few non EU countries restricted the use of certain neonicotinoids.

Organophosphate and carbamate insecticides have a similar mode of action. They affect the nervous system of target pests (and non-target organisms) by disrupting acetylcholinesterase activity, the enzyme that regulates acetylcholine, at nerve synapses. This inhibition causes an increase in synaptic acetylcholine and over-stimulation of the parasympathetic nervous system. Many of these insecticides, first developed in the mid 20th century, are very poisonous. Although commonly used in the past, many older chemicals have been removed from the market due to their health and environmental effects (*e.g.* DDT, chlordane, and toxaphene). However, many organophosphates are not persistent in the environment.

Pyrethroid insecticides were developed as a synthetic version of the naturally occurring pesticide pyrethrin, which is found in chrysanthemums. They have been modified to increase their stability in the environment. Some synthetic pyrethroids are toxic to the nervous system.

Herbicides

A number of sulfonylureas have been commercialized for weed control, including: amidosulfuron, flazasulfuron, metsulfuron-methyl, rimsulfuron, sulfometuron-methyl, terbacil, nicosulfuron, and triflusulfuron-methyl. These are broad-spectrum herbicides that kill plants weeds or pests by inhibiting the enzyme acetolactate synthase. In the 1960s, more than 1 kg/ha (0.89 lb/acre) crop protection chemical was typically applied, while sulfonylureates allow as little as 1% as much material to achieve the same effect.

Biopesticides

Biopesticides are certain types of pesticides derived from such natural materials as animals, plants, bacteria, and certain minerals. For example, canola oil and baking soda have pesticidal applications and are considered biopesticides. Biopesticides fall into three major classes:

- Microbial pesticides which consist of bacteria, entomopathogenic fungi or viruses (and sometimes includes the metabolites that bacteria or fungi produce). Entomopathogenic nematodes are also often classed as microbial pesticides, even though they are multi-cellular.
- Biochemical pesticides or herbal pesticides are naturally occurring substances that control (or monitor in the case of pheromones) pests and microbial diseases.
- Plant-incorporated protectants (PIPs) have genetic material from other species incorporated into their genetic material (i.e. GM crops). Their use is controversial, especially in many European countries.

Classified by Type of Pest

Pesticides that are related to the type of pests are:

Type	Action
Algicides	Control algae in lakes, canals, swimming pools, water tanks, and other sites
Antifouling agents	Kill or repel organisms that attach to underwater surfaces, such as boat bottoms
Antimicrobials	Kill microorganisms (such as bacteria and viruses)
Attractants	Attract pests (for example, to lure an insect or rodent to a trap). (However, food is not considered a pesticide when used as an attractant.)
Biopesticides	Biopesticides are certain types of pesticides derived from such natural materials as animals, plants, bacteria, and certain minerals
Biocides	Kill microorganisms
Disinfectants and sanitizers	Kill or inactivate disease-producing microorganisms on inanimate objects
Fungicides	Kill fungi (including blights, mildews, molds, and rusts)
Fumigants	Produce gas or vapor intended to destroy pests in buildings or soil
Herbicides	Kill weeds and other plants that grow where they are not wanted
Insecticides	Kill insects and other arthropods
Miticides	Kill mites that feed on plants and animals

Microbial pesticides	Microorganisms that kill, inhibit, or out compete pests, including insects or other microorganisms
Molluscicides	Kill snails and slugs
Nematicides	Kill nematodes (microscopic, worm-like organisms that feed on plant roots)
Ovicides	Kill eggs of insects and mites
Pheromones	Biochemicals used to disrupt the mating behavior of insects
Repellents	Repel pests, including insects (such as mosquitoes) and birds
Rodenticides	Control mice and other rodents

Further Types

The term pesticide also include these substances:

Defoliants: Cause leaves or other foliage to drop from a plant, usually to facilitate harvest.

Desiccants: Promote drying of living tissues, such as unwanted plant tops.

Insect growth regulators: Disrupt the molting, maturity from pupal stage to adult, or other life processes of insects.

Plant growth regulators: Substances (excluding fertilizers or other plant nutrients) that alter the expected growth, flowering, or reproduction rate of plants.

Wood preservatives: They are used to make wood resistant to insects, fungus, and other pests.

Residue

Pesticide residue refers to the pesticides that may remain on or in food after they are applied to food crops. The maximum allowable levels of these residues in foods are often stipulated by regulatory bodies in many countries. Regulations such as pre-harvest intervals also often prevent harvest of crop or livestock products if recently treated in order to allow residue concentrations to decrease over time to safe levels before harvest. Exposure of the general population to these residues most commonly occurs through consumption of treated food sources, or being in close contact to areas treated with pesticides such as farms or lawns.

Many of these chemical residues, especially derivatives of chlorinated pesticides, exhibit bioaccumulation which could build up to harmful levels in the body as well as in the environment. Persistent chemicals can be magnified through the food chain and have been detected in products ranging from meat, poultry, and fish, to vegetable oils, nuts, and various fruits and vegetables.

Because of the toxicity of pesticides in concentrated form and because of the frequency of exposure, the most serious human health effects are found among agricultural and production workers. Long-term and chronic health effects occur as the chemicals are ingested and inhaled. Bioaccumulation (the buildup of toxic materials in tissues) is evident in fish and birds as well as in humans.

Two different sets of problems are posed by the occurrence of persistent pesticides in the environment: (1) localized problems, tending to be acute, leading to recognizable effects with assignable origins, which can be dealt with if successfully sorted out and if the will and powers exist; and (2) more widespread problems, tending to be inferential with postulated effects, due to the universal presence of the materials in question. There is, for example, growing evidence from the amount of pesticide residues found in specimens of affected species as well as in the animals, fish, invertebrates, or plankton they feed upon, that there is a process of pesticide concentration in the food chains. As small amounts of persistent pesticides become more and more widely spread throughout the entire natural environment, they are absorbed by low forms of life. Where large numbers of these species serve as food for higher animals, some of the total pesticide remains in the eater. This can lead to a certain concentration of pesticides in a form of life at the end of the food chain. Eventually lethal doses may be reached for certain populations, or the species may be reduced because of adverse effects on reproduction or behaviour.

The importance of changes in the environment is that they may be irreversible: species may disappear altogether. This will cause ecological imbalance - certain forms of life whose numbers are controlled by the disappearing species will no longer be held in check. Thus an apparently harmless species may become a pest. Another result of reduction of species diversity may be the loss of the genetic possibilities that each organism disturbance of the selective pressures of the environment will reduce the chances of future development of desired plants and animals. Furthermore, as humans are at the top of food chains, they too tend to concentrate residues in their bodies with as yet unknown effects on them. While perhaps sufficient attention has been directed to acute toxicity problems, too little attention has been paid to the effect of long term ingestion by human beings of small amounts of these chemicals.

A serious dilemma is that the cheapest pesticides are the organochlorines, known to accumulate in human fat. The substitution of organophosphorus pesticides brings both higher cost and a need for more carefully trained operators, both of which pose problems.

Herbicide

A herbicide is a chemical substance used to control or manipulate undesirable vegetation, especially weeds. Herbicides are extensively used in gardening, farming, and landscape turf management.

Herbicides are classified into two categories: selective and non-selective. Selective herbicides kill specific unwanted plants while leaving desirable vegetation relatively unharmed. Non-selective herbicides (total weed killers) kill all or most plant species.

A herbicide can be applied directly to the plant, applied to the soil, or sprayed onto the foliage. Herbicides are applied before, during, or after crop planting in row-crop farming to maximize crop production by diminishing the development of unwanted plants. Herbicides are also applied in ponds and lakes to control aquatic plants, in forests to prepare logged areas for replanting, and to golf courses, lawns, parks, and other areas to clear out unwanted vegetation.

Besides selective and non-selective classifications, a herbicide can also be categorized according to three other characteristics:

1. Persistence - How long the herbicide remains potent
2. Mechanism of action - How it works
3. Means of uptake - How the plants will absorb it (e.g., through the roots, aboveground foliage, etc.)

A herbicide's effectiveness is strongly influenced by its toxic mode of action and the application method. Herbicides can act by inhibiting a plant's amino acid production, growth, photosynthesis, cell division, or by mimicking natural auxin hormones to cause deformities. Most modern herbicides are synthetic mimics of a natural plant's hormones that obstruct the target plant's growth.

Some plants such as the tree of heaven and juglans (walnuts) produce their own natural herbicide. Organic herbicides are useful and are commonly used in organic gardens, but they are less effective and more costly than synthetic herbicides because they based on natural materials.

For difficult cases, a combination of several herbicides is recommended when dealing with herbicide resistance.

Terminology

Herbicides are classified/grouped in various ways e.g. according to the activity, timing of application, method of application, mechanism of action, chemical family. This gives rise to a considerable level of terminology related to herbicides and their use.

Intended Outcome

- Control is the destruction of unwanted weeds, or the damage of them to the point where they are no longer competitive with the crop.

- Suppression is incomplete control still providing some economic benefit, such as reduced competition with the crop.
- Crop safety, for selective herbicides, is the relative absence of damage or stress to the crop. Most selective herbicides cause some visible stress to crop plants.
- Defoliant, similar to herbicides, but designed to remove foliage (leaves) rather than kill the plant.

Selectivity

- Selective herbicides control or suppress certain plants without affecting the growth of other plants species. Selectivity may be due to translocation, differential absorption, physical (morphological) or physiological differences between plant species. 2,4-D, mecoprop, dicamba control many broadleaf weeds but remain ineffective against turfgrasses.
- Non-selective herbicides are not specific in acting against certain plant species and control all plant material with which they come into contact. They are used to clear industrial sites, waste ground, railways and railway embankments. Paraquat, glufosinate, glyphosate are non-selective herbicides.

Timing of Application

- Preplant: Preplant herbicides are nonselective herbicides applied to soil before planting. Some preplant herbicides may be mechanically incorporated into the soil. The objective for incorporation is to prevent dissipation through photodecomposition and/or volatility. The herbicides kill weeds as they grow through the herbicide treated zone. Volatile herbicides have to be incorporated into the soil before planting the pasture. Agricultural crops grown in soil treated with a preplant herbicide include tomatoes, corn, soybeans and strawberries. Soil fumigants like metam-sodium and dazomet are in use as preplant herbicides.
- Preemergence: Preemergence herbicides are applied before the weed seedlings emerge through the soil surface. Herbicides do not prevent weeds from germinating but they kill weeds as they grow through the herbicide treated zone by affecting the cell division in the emerging seedling. Dithopyr and pendimethalin are preemergence herbicides. Weeds that have already emerged before application or activation are not affected by pre-herbicides as their primary growing point escapes the treatment.
- Postemergence: These herbicides are applied after weed seedlings have emerged through the soil surface. They can be foliar or root absorbed, selective or nonselective, contact or systemic. Application of these herbicides is avoided during rain because the problem of being washed off to the soil makes it ineffective. 2,4-D is a selective, systemic, foliar absorbed postemergence herbicide.

Method of Application

- Soil applied: Herbicides applied to the soil are usually taken up by the root or shoot of the emerging seedlings and are used as preplant or preemergence treatment. Several factors influence the effectiveness of soil-applied herbicides. Weeds absorb herbicides by both passive and active mechanism. Herbicide adsorption to soil colloids or organic matter often reduces its amount available for weed absorption. Positioning of herbicide in correct layer of soil is very important, which can be achieved mechanically and by rainfall. Herbicides on the soil surface are subjected to several processes that reduce their availability. Volatility and photolysis are two common processes that reduce the availability of herbicides. Many soil applied herbicides are absorbed through plant shoots while they are still underground leading to their death or injury. EPTC and trifluralin are soil applied herbicides.
- Foliar applied: These are applied to portion of the plant above the ground and are absorbed by exposed tissues. These are generally postemergence herbicides and can either be translocated (systemic) throughout the plant or remain at specific site (contact). External barriers of plants like cuticle, waxes, cell wall etc. affect herbicide absorption and action. Glyphosate, 2,4-D and dicamba are foliar applied herbicide.

Persistence

- Residual activity: An herbicide is described as having low residual activity if it is neutralized within a short time of application (within a few weeks or months) - typically this is due to rainfall, or by reactions in the soil. An herbicide described as having high residual activity will remain potent for a long term in the soil. For some compounds, the residual activity can leave the ground almost permanently barren.

Mechanism of Action

Herbicides are often classified according to their site of action, because as a general rule, herbicides within the same site of action class will produce similar symptoms on susceptible plants. Classification based on site of action of herbicide is comparatively better as herbicide resistance management can be handled more properly and effectively. Classification by mechanism of action (MOA) indicates the first enzyme, protein, or biochemical step affected in the plant following application.

List of Mechanisms found in Modern Herbicides

- ACCase inhibitors: Acetyl coenzyme A carboxylase (ACCase) is part of the first step of lipid synthesis. Thus, ACCase inhibitors affect cell membrane production in the meristems of the grass plant. The ACCases of grasses are sensitive to these herbicides, whereas the ACCases of dicot plants are not.

- ALS inhibitors: the acetolactate synthase (ALS) enzyme (also known as acetohydroxyacid synthase, or AHAS) is the first step in the synthesis of the branched-chain amino acids (valine, leucine, and isoleucine). These herbicides slowly starve affected plants of these amino acids, which eventually leads to inhibition of DNA synthesis. They affect grasses and dicots alike. The ALS inhibitor family includes various sulfonylureas (SUs) (such as Flazasulfuron and Metsulfuron-methyl), imidazolinones (IMIs), triazolopyrimidines (TPs), pyrimidinyl oxybenzoates (POBs), and sulfonylamino carbonyl triazolinones(SCTs). The ALS biological pathway exists only in plants and not animals, thus making the ALS-inhibitors among the safest herbicides.
- EPSPS inhibitors: Enolpyruvylshikimate 3-phosphate synthase enzyme (EPSPS) is used in the synthesis of the amino acids tryptophan, phenylalanine and tyrosine. They affect grasses and dicots alike. Glyphosate (Roundup) is a systemic EPSPS inhibitor inactivated by soil contact.
- Synthetic auxins inaugurated the era of organic herbicides. They were discovered in the 1940s after a long study of the plant growth regulator auxin. Synthetic auxins mimic this plant hormone. They have several points of action on the cell membrane, and are effective in the control of dicot plants. 2,4-D is a synthetic auxin herbicide.
- Photosystem II inhibitors reduce electron flow from water to $NADP^+$ at the photochemical step in photosynthesis. They bind to the Qb site on the D1 protein, and prevent quinone from binding to this site. Therefore, this group of compounds causes electrons to accumulate on chlorophyll molecules. As a consequence, oxidation reactions in excess of those normally tolerated by the cell occur, and the plant dies. The triazine herbicides (including atrazine) and urea derivatives (diuron) are photosystem II inhibitors.
- Photosystem I inhibitors steal electrons from the normal pathway through FeS to Fdx to $NADP^+$ leading to direct discharge of electrons on oxygen. As a result, reactive oxygen species are produced and oxidation reactions in excess of those normally tolerated by the cell occur, leading to plant death. Bipyridinium herbicides (such as diquat and paraquat) inhibit the FeS to Fdx step of that chain, while diphenyl ether herbicides (such as nitrofen, nitrofluorfen, and acifluorfen) inhibit the Fdx to $NADP^+$ step.
- HPPD inhibitors inhibit 4-Hydroxyphenylpyruvate dioxygenase, which are involved in tyrosine breakdown. Tyrosine breakdown products are used by plants to make carotenoids, which protect chlorophyll in plants from being destroyed by sunlight. If this happens, the plants turn white due to complete loss of chlorophyll, and the plants die. Mesotrione and sulcotrione are herbicides in this class; a drug, nitisinone, was discovered in the course of developing this class of herbicides.

Herbicide Group

One of the most important methods for preventing, delaying, or managing resistance is to reduce the reliance on a single herbicide mode of action. To do this, farmers must know the mode of action for the herbicides they intend to use, but the relatively complex nature of plant biochemistry makes this difficult to determine. Attempts were made to simplify the understanding of herbicide mode of action by developing a classification system that grouped herbicides by mode of action. Eventually the Herbicide Resistance Action Committee (HRAC) and the Weed Science Society of America (WSSA) developed a classification system. The WSSA and HRAC systems differ in the group designation. Groups in the WSSA and the HRAC systems are designated by numbers and letters, respectively. The goal for adding the "Group" classification and mode of action to the herbicide product label is to provide a simple and practical approach to deliver the information to users. This information will make it easier to develop educational material that is consistent and effective. It should increase user's awareness of herbicide mode of action and provide more accurate recommendations for resistance management.Another goal is to make it easier for users to keep records on which herbicide mode of actions are being used on a particular field from year to year.

Chemical Family

Detailed investigations on chemical structure of the active ingredients of the registered herbicides showed that some moieties (moiety is a part of a molecule that may include either whole functional groups or parts of functional groups as substructures; a functional group has similar chemical properties whenever it occurs in different compounds) have the same mechanisms of action. According to Forouzesh *et al.* 2015, these moieties have been assigned to the names of chemical families and active ingredients are then classified within the chemical families accordingly. Knowing about herbicide chemical family grouping could serve as a short-term strategy for managing resistance to site of action.

Use and Application

Herbicides being sprayed from the spray arms of a tractor in North Dakota.

Most herbicides are applied as water-based sprays using ground equipment. Ground equipment varies in design, but large areas can be sprayed using self-propelled sprayers equipped with long booms, of 60 to 120 feet (18 to 37 m) with spray nozzles spaced every 20–30 inches (510–760 mm) apart. Towed, handheld, and even horse-drawn sprayers are also used. On large areas, herbicides may also at times be applied aerially using helicopters or airplanes, or through irrigation systems (known as chemigation).

A further method of herbicide application developed around 2010, involves ridding the soil of its active weed seed bank rather than just killing the weed. This can successfully treat annual plants but not perennials. Researchers at the Agricultural Research Service found that the application of herbicides to fields late in the weeds› growing season greatly reduces their seed production, and therefore fewer weeds will return the following season. Because most weeds are annuals, their seeds will only survive in soil for a year or two, so this method will be able to destroy such weeds after a few years of herbicide application.

Weed-wiping may also be used, where a wick wetted with herbicide is suspended from a boom and dragged or rolled across the tops of the taller weed plants. This allows treatment of taller grassland weeds by direct contact without affecting related but desirable shorter plants in the grassland sward beneath. The method has the benefit of avoiding spray drift. In Wales, a scheme offering free weed-wiper hire was launched in 2015 in an effort to reduce the levels of MCPA in water courses.

Misuse and Misapplication

Herbicide volatilisation or spray drift may result in herbicide affecting neighboring fields or plants, particularly in windy conditions. Sometimes, the wrong field or plants may be sprayed due to error.

Health and Environmental Effects

Herbicides have widely variable toxicity in addition to acute toxicity arising from ingestion of a significant quantity rapidly, and chronic toxicity arising from environmental and occupational exposure over long periods. Much public suspicion of herbicides revolves around a confusion between valid statements of *acute* toxicity as opposed to equally valid statements of lack of *chronic* toxicity at the recommended levels of usage. For instance, while glyphosate formulations with talloamine *adjuvants* are acutely toxic, their use was found to be uncorrelated with any health issues like cancer in a massive US Department of Health study on 90,000 members of farmer families for over a period of 23 years . That is, the study shows lack of chronic toxicity, but cannot question the herbicide's acute toxicity.

Some herbicides cause a range of health effects ranging from skin rashes to death. The pathway of attack can arise from intentional or unintentional direct consumption, improper application resulting in the herbicide coming into direct contact with people or wildlife, inhalation of aerial sprays, or food consumption prior to the labelled preharvest interval. Under some conditions, certain herbicides can be transported via leaching or surface runoff to contaminate groundwater or distant surface water sources. Generally, the conditions that promote herbicide transport include intense storm events (particularly shortly after application) and soils with limited capacity to adsorb or retain the herbicides. Herbicide properties that increase likelihood of transport include persistence (resistance to degradation) and high water solubility.

Phenoxy herbicides are often contaminated with dioxins such as TCDD; research has suggested such contamination results in a small rise in cancer risk after occupational exposure to these herbicides. Triazine exposure has been implicated in a likely relationship to increased risk of breast cancer, although a causal relationship remains unclear.

Herbicide manufacturers have at times made false or misleading claims about the safety of their products. Chemical manufacturer Monsanto Company agreed to change its advertising after pressure from New York attorney general Dennis Vacco; Vacco complained about misleading claims that its spray-on glyphosate-based herbicides, including Roundup, were safer than table salt and "practically non-toxic" to mammals, birds, and fish (though proof that this was ever said is hard to find). Roundup is toxic and has resulted in death after being ingested in quantities ranging from 85 to 200 ml, although it has also been ingested in quantities as large as 500 ml with only mild or moderate symptoms. The manufacturer of Tordon 101 (Dow AgroSciences, owned by the Dow Chemical Company) has claimed Tordon 101 has no effects on animals and insects, in spite of evidence of strong carcinogenic activity of the active ingredient Picloram in studies on rats.

The risk of Parkinson's disease has been shown to increase with occupational exposure to herbicides and pesticides. The herbicide paraquat is suspected to be one such factor.

All commercially sold, organic and nonorganic herbicides must be extensively tested prior to approval for sale and labeling by the Environmental Protection Agency. However, because of the large number of herbicides in use, concern regarding health effects is significant. In addition to health effects caused by herbicides themselves, commercial herbicide mixtures often contain other chemicals, including inactive ingredients, which have negative impacts on human health.

Ecological Effects

Commercial herbicide use generally has negative impacts on bird populations, although the impacts are highly variable and often require field studies to predict accurately. Laboratory studies have at times overestimated negative impacts on birds due to toxicity, predicting serious problems that were not observed in the field. Most observed effects are due not to toxicity, but to habitat changes and the decreases in abundance of species on which birds rely for food or shelter. Herbicide use in silviculture, used to favor certain types of growth following clearcutting, can cause significant drops in bird populations. Even when herbicides which have low toxicity to birds are used, they decrease the abundance of many types of vegetation on which the birds rely. Herbicide use in agriculture in Britain has been linked to a decline in seed-eating bird species which rely on the weeds killed by the herbicides. Heavy use of herbicides in neotropical agricultural areas has been one of many factors implicated in limiting the usefulness of such agricultural land for wintering migratory birds.

Frog populations may be affected negatively by the use of herbicides as well. While some studies have shown that atrazine may be a teratogen, causing demasculinization in male frogs, the U.S. Environmental Protection Agency (EPA) and its independent Scientific Advisory Panel (SAP) examined all available studies on this topic and concluded that "atrazine does not adversely affect amphibian gonadal development based on a review of laboratory and field studies."

Scientific Uncertainty of Full Extent of Herbicide Effects

The health and environmental effects of many herbicides is unknown, and even the scientific community often disagrees on the risk. For example, a 1995 panel of 13 scientists reviewing studies on the carcinogenicity of 2,4-D had divided opinions on the likelihood 2,4-D causes cancer in humans. As of 1992, studies on phenoxy herbicides were too few to accurately assess the risk of many types of cancer from these herbicides, even though evidence was stronger that exposure to these herbicides is associated with increased risk of soft tissue sarcoma and non-Hodgkin lymphoma. Furthermore, there is some suggestion that herbicides can play a role in sex reversal of certain organisms that experience temperature-dependent sex determination, which could theoretically alter sex ratios.

Resistance

Weed resistance to herbicides has become a major concern in crop production worldwide. Resistance to herbicides is often attributed to lack of rotational programmes of herbicides and to continuous applications of herbicides with the same sites of action. Thus, a true understanding of the sites of action of herbicides is essential for strategic planning of herbicide-based weed control.

Plants have developed resistance to atrazine and to ALS-inhibitors, and more recently, to glyphosate herbicides. Marestail is one weed that has developed glyphosate resistance. Glyphosate-resistant weeds are present in the vast majority of soybean, cotton and corn farms in some U.S. states. Weeds that can resist multiple other herbicides are spreading. Few new herbicides are near commercialization, and none with a molecular mode of action for which there is no resistance. Because most herbicides could not kill all weeds, farmers rotated crops and herbicides to stop resistant weeds. During its initial years, glyphosate was not subject to resistance and allowed farmers to reduce the use of rotation.

A family of weeds that includes waterhemp (Amaranthus rudis) is the largest concern. A 2008-9 survey of 144 populations of waterhemp in 41 Missouri counties revealed glyphosate resistance in 69%. Weeds from some 500 sites throughout Iowa in 2011 and 2012 revealed glyphosate resistance in approximately 64% of waterhemp samples. The use of other killers to target "residual" weeds has become common, and may be sufficient to have stopped the spread of resistance From 2005 through 2010 researchers discovered 13 different weed species that had developed resistance to glyphosate. But since then only two more have been discovered. Weeds resistant to multiple herbicides with completely different biological action modes are on the rise. In Missouri, 43% of samples were resistant to two different herbicides; 6% resisted three; and 0.5% resisted four. In Iowa 89% of waterhemp samples resist two or more herbicides, 25% resist three, and 10% resist five.

For southern cotton, herbicide costs has climbed from between $50 and $75 per hectare a few years ago to about $370 per hectare in 2013. Resistance is contributing to a massive shift away from growing cotton; over the past few years, the area planted with cotton has declined by 70% in Arkansas and by 60% in Tennessee. For soybeans in Illinois, costs have risen from about $25 to $160 per hectare.

Biochemistry of Resistance

Resistance to herbicides can be based on one of the following biochemical mechanisms:

- Target-site resistance: This is due to a reduced (or even lost) ability of the herbicide to bind to its target protein. The effect usually relates to an enzyme with a crucial function in a metabolic pathway, or to a component of an electron-transport system. Target-site resistance may also be caused by an overexpression of the target enzyme (via gene amplification or changes in a gene promoter).
- Non-target-site resistance: This is caused by mechanisms that reduce the amount of herbicidal active compound reaching the target site. One important mechanism is an enhanced metabolic detoxification of the herbicide in the weed, which leads to insufficient amounts of the active substance reaching the target site. A reduced uptake and translocation, or sequestration of the herbicide, may also result in an insufficient herbicide transport to the target site.
- Cross-resistance: In this case, a single resistance mechanism causes resistance to several herbicides. The term target-site cross-resistance is used when the herbicides bind to the same target site, whereas non-target-site cross-resistance is due to a single non-target-site mechanism (e.g., enhanced metabolic detoxification) that entails resistance across herbicides with different sites of action.

- Multiple resistance: In this situation, two or more resistance mechanisms are present within individual plants, or within a plant population.

Resistance Management

Worldwide experience has been that farmers tend to do little to prevent herbicide resistance developing, and only take action when it is a problem on their own farm or neighbor's. Careful observation is important so that any reduction in herbicide efficacy can be detected. This may indicate evolving resistance. It is vital that resistance is detected at an early stage as if it becomes an acute, whole-farm problem, options are more limited and greater expense is almost inevitable. Table below lists factors which enable the risk of resistance to be assessed. An essential pre-requisite for confirmation of resistance is a good diagnostic test. Ideally this should be rapid, accurate, cheap and accessible. Many diagnostic tests have been developed, including glasshouse pot assays, petri dish assays and chlorophyll fluorescence. A key component of such tests is that the response of the suspect population to a herbicide can be compared with that of known susceptible and resistant standards under controlled conditions. Most cases of herbicide resistance are a consequence of the repeated use of herbicides, often in association with crop monoculture and reduced cultivation practices. It is necessary, therefore, to modify these practices in order to prevent or delay the onset of resistance or to control existing resistant populations. A key objective should be the reduction in selection pressure. An integrated weed management (IWM) approach is required, in which as many tactics as possible are used to combat weeds. In this way, less reliance is placed on herbicides and so selection pressure should be reduced.

Optimising herbicide input to the economic threshold level should avoid the unnecessary use of herbicides and reduce selection pressure. Herbicides should be used to their greatest potential by ensuring that the timing, dose, application method, soil and climatic conditions are optimal for good activity. In the UK, partially resistant grass weeds such as *Alopecurus myosuroides* (blackgrass) and *Avena* spp. (wild oat) can often be controlled adequately when herbicides are applied at the 2-3 leaf stage, whereas later applications at the 2-3 tiller stage can fail badly. Patch spraying, or applying herbicide to only the badly infested areas of fields, is another means of reducing total herbicide use.

Table: Agronomic factors influencing the risk of herbicide resistance development

Factor	Low risk	High risk
Cropping system	Good rotation	Crop monoculture
Cultivation system	Annual ploughing	Continuous minimum tillage
Weed control	Cultural only	Herbicide only
Herbicide use	Many modes of action	Single modes of action
Control in previous years	Excellent	Poor
Weed infestation	Low	High
Resistance in vicinity	Unknown	Common

Approaches to Treating Resistant Weeds

Alternative Herbicides

When resistance is first suspected or confirmed, the efficacy of alternatives is likely to be the first consideration. The use of alternative herbicides which remain effective on resistant populations can be a successful strategy, at least in the short term. The effectiveness of alternative herbicides will be highly dependent on the extent of cross-resistance. If there is resistance to a single group of herbicides, then the use of herbicides from other groups may provide a simple and effective solution, at least in the short term. For example, many triazine-resistant weeds have been readily controlled by the use of alternative herbicides such as dicamba or glyphosate. If resistance extends to more than one herbicide group, then choices are more limited. It should not be assumed that resistance will automatically extend to all herbicides with the same mode of action, although it is wise to assume this until proved otherwise. In many weeds the degree of cross-resistance between the five groups of ALS inhibitors varies considerably. Much will depend on the resistance mechanisms present, and it should not be assumed that these will necessarily be the same in different populations of the same species. These differences are due, at least in part, to the existence of different mutations conferring target site resistance. Consequently, selection for different mutations may result in different patterns of cross-resistance. Enhanced metabolism can affect even closely related herbicides to differing degrees. For example, populations of *Alopecurus myosuroides* (blackgrass) with an enhanced metabolism mechanism show resistance to pendimethalin but not to trifluralin, despite both being dinitroanilines. This is due to differences in the vulnerability of these two herbicides to oxidative metabolism. Consequently, care is needed when trying to predict the efficacy of alternative herbicides.

Mixtures and Sequences

The use of two or more herbicides which have differing modes of action can reduce the selection for resistant genotypes. Ideally, each component in a mixture should:

- Be active at different target sites
- Have a high level of efficacy
- Be detoxified by different biochemical pathways
- Have similar persistence in the soil (if it is a residual herbicide)
- Exert negative cross-resistance
- Synergise the activity of the other component

No mixture is likely to have all these attributes, but the first two listed are the most important. There is a risk that mixtures will select for resistance to both components in the longer term. One practical advantage of sequences of two herbicides compared with mixtures is that a better appraisal of the efficacy of each herbicide component is possible, provided that sufficient time elapses between each application. A disadvantage with sequences is that two separate applications have to be made and it is possible that the later application will be less effective on weeds surviving the first application. If these are resistant, then the second herbicide in the sequence may increase selection for resistant individuals by killing the susceptible plants which were damaged but not killed by the first application, but allowing the larger, less affected, resistant plants to survive. This has been cited as one reason why ALS-resistant *Stellaria media* has evolved in Scotland recently (2000), despite the regular use of a sequence incorporating mecoprop, a herbicide with a different mode of action.

Herbicide Rotations

Rotation of herbicides from different chemical groups in successive years should reduce selection for resistance. This is a key element in most resistance prevention programmes. The value of this approach depends on the extent of cross-resistance, and whether multiple resistance occurs owing to the presence of several different resistance mechanisms. A practical problem can be the lack of awareness by farmers of the different groups of herbicides that exist. In Australia a scheme has been introduced in which identifying letters are included on the product label as a means of enabling farmers to distinguish products with different modes of action.

Farming Practices and Resistance

Herbicide resistance became a critical problem in Australian agriculture, after many Australian sheep farmers began to exclusively grow wheat in their pastures in the 1970s. Introduced varieties of ryegrass, while good for grazing sheep, compete intensely with wheat. Ryegrasses produce so many seeds that, if left unchecked, they can completely choke a field. Herbicides provided excellent control, while reducing soil disrupting because of less need to plough. Within little more than a decade, ryegrass and other weeds began to develop resistance. In response Australian farmers changed methods. By 1983, patches of ryegrass had become immune to Hoegrass, a family of herbicides that inhibit an enzyme called acetyl coenzyme A carboxylase.

Ryegrass populations were large, and had substantial genetic diversity, because farmers had planted many varieties. Ryegrass is cross-pollinated by wind, so genes shuffle frequently. To control its distribution farmers sprayed inexpensive Hoegrass, creating selection pressure. In addition, farmers sometimes diluted the herbicide in order to save money, which allowed some plants to survive application. When resistance appeared farmers turned to a group of herbicides that block acetolactate synthase. Once again, ryegrass in Australia evolved a kind of "cross-resistance" that allowed it to rapidly break down a variety of herbicides. Four classes of herbicides become ineffective within a few years. In 2013 only two herbicide classes, called Photosystem II and long-chain fatty acid inhibitors, were effective against ryegrass.

List of Common Herbicides

Synthetic Herbicides

- 2,4-D is a broadleaf herbicide in the phenoxy group used in turf and no-till field crop production. Now, it is mainly used in a blend with other herbicides to allow lower rates of herbicides to be used; it is the most widely used herbicide in the world, and third most commonly used in the United States. It is an example of synthetic auxin (plant hormone).
- Aminopyralid is a broadleaf herbicide in the pyridine group, used to control weeds on grassland, such as docks, thistles and nettles. It is notorious for its ability to persist in compost.
- Atrazine, a triazine herbicide, is used in corn and sorghum for control of broadleaf weeds and grasses. Still used because of its low cost and because it works well on a broad spectrum of weeds common in the US corn belt, atrazine is commonly used with other herbicides to reduce the overall rate of atrazine and to lower the potential for groundwater contamination; it is a photosystem II inhibitor.
- Clopyralid is a broadleaf herbicide in the pyridine group, used mainly in turf, rangeland, and for control of noxious thistles. Notorious for its ability to persist in compost, it is another example of synthetic auxin.
- Dicamba, a postemergent broadleaf herbicide with some soil activity, is used on turf and field corn. It is another example of a synthetic auxin.
- Glufosinate ammonium, a broad-spectrum contact herbicide, is used to control weeds after the crop emerges or for total vegetation control on land not used for cultivation.
- Fluazifop (Fuselade Forte), a post emergence, foliar absorbed, translocated grass-selective herbicide with little residual action. It is used on a very wide range of broad leaved crops for control of annual and perennial grasses.
- Fluroxypyr, a systemic, selective herbicide, is used for the control of broad-leaved weeds in small grain cereals, maize, pastures, rangeland and turf. It is a synthetic auxin. In cereal growing, fluroxypyr's key importance is control of cleavers, Galium aparine. Other key broadleaf weeds are also controlled.
- Glyphosate, a systemic nonselective herbicide, is used in no-till burndown and for weed control in crops genetically modified to resist its effects. It is an example of an EPSPs inhibitor.
- Imazapyr a nonselective herbicide, is used for the control of a broad range of weeds, including terrestrial annual and perennial grasses and broadleaf herbs, woody species, and riparian and emergent aquatic species.

- Imazapic, a selective herbicide for both the pre- and postemergent control of some annual and perennial grasses and some broadleaf weeds, kills plants by inhibiting the production of branched chain amino acids (valine, leucine, and isoleucine), which are necessary for protein synthesis and cell growth.
- Imazamox, an imidazolinone manufactured by BASF for postemergence application that is an acetolactate synthase (ALS) inhibitor. Sold under trade names Raptor, Beyond, and Clearcast.
- Linuron is a nonselective herbicide used in the control of grasses and broadleaf weeds. It works by inhibiting photosynthesis.
- MCPA (2-methyl-4-chlorophenoxyacetic acid) is a phenoxy herbicide selective for broadleaf plants and widely used in cereals and pasture.
- Metolachlor is a pre-emergent herbicide widely used for control of annual grasses in corn and sorghum; it has displaced some of the atrazine in these uses.
- Paraquat is a nonselective contact herbicide used for no-till burndown and in aerial destruction of marijuana and coca plantings. It is more acutely toxic to people than any other herbicide in widespread commercial use.
- Pendimethalin, a pre-emergent herbicide, is widely used to control annual grasses and some broad-leaf weeds in a wide range of crops, including corn, soybeans, wheat, cotton, many tree and vine crops, and many turfgrass species.
- Picloram, a pyridine herbicide, mainly is used to control unwanted trees in pastures and edges of fields. It is another synthetic auxin.
- Sodium chlorate (disused/banned in some countries), a nonselective herbicide, is considered phytotoxic to all green plant parts. It can also kill through root absorption.
- Triclopyr, a systemic, foliar herbicide in the pyridine group, is used to control broadleaf weeds while leaving grasses and conifers unaffected.
- Several sulfonylureas, including Flazasulfuron and Metsulfuron-methyl, which act as ALS inhibitors and in some cases are taken up from the soil via the roots.

Organic Herbicides

Recently, the term "organic" has come to imply products used in organic farming. Under this definition, an organic herbicide is one that can be used in a farming enterprise that has been classified as organic. Depending on the application, they may be less effective than synthetic herbicides and are generally used along with cultural and mechanical weed control practices.

Homemade organic herbicides include:

- Corn gluten meal (CGM) is a natural pre-emergence weed control used in turfgrass, which reduces germination of many broadleaf and grass weeds.
- Vinegar is effective for 5–20% solutions of acetic acid, with higher concentrations most effective, but it mainly destroys surface growth, so respraying to treat regrowth is needed. Resistant plants generally succumb when weakened by respraying.
- Steam has been applied commercially, but is now considered uneconomical and inadequate. It controls surface growth but not underground growth and so respraying to treat regrowth of perennials is needed.
- Flame is considered more effective than steam, but suffers from the same difficulties.
- D-limonene (citrus oil) is a natural degreasing agent that strips the waxy skin or cuticle from weeds, causing dehydration and ultimately death.
- Saltwater or salt applied in appropriate strengths to the rootzone will kill most plants.
- Monocerin produced by certain fungi will kill certain weeds such as Johnson grass.

Other Herbicides

- 2,4,5-Trichlorophenoxyacetic acid (2,4,5-T) was a widely used broadleaf herbicide until being phased out starting in the late 1970s. While 2,4,5-T itself is of only moderate toxicity, the manufacturing process for 2,4,5-T contaminates this chemical with trace amounts of 2,3,7,8-tetrachlorodibenzo-p-dioxin (TCDD). TCDD is extremely toxic to humans. With proper temperature control during production of 2,4,5-T, TCDD levels can be held to about .005 ppm. Before the TCDD risk was well understood, early production facilities lacked proper temperature controls. Individual batches tested later were found to have as much as 60 ppm of TCDD. 2,4,5-T was withdrawn from use in the USA in 1983, at a time of heightened public sensitivity about chemical hazards in the environment. Public concern about dioxins was high, and production and use of other (non-herbicide) chemicals potentially containing TCDD contamination was also withdrawn. These included pentachlorophenol (a wood preservative) and PCBs (mainly used as stabilizing agents in transformer oil). Some feel that the 2,4,5-T withdrawal was not based on sound science. 2,4,5-T has since largely been replaced by dicamba and triclopyr.
- Agent Orange was a herbicide blend used by the British military during the Malayan Emergency and the U.S. military during the Vietnam War between January 1965 and April 1970 as a defoliant. It was a 50/50 mixture of the *n*-butyl esters of 2,4,5-T and 2,4-D. Because of TCDD contamination in the 2,4,5-T component, it has been blamed for serious illnesses in many people who were exposed to it.

- Diesel, and other heavy oil derivatives, are known to be informally used at times, but are usually banned for this purpose.

Molluscicide

Molluscicides are pesticides which kill mollusks, an animal phylum of tens of thousands of invertebrate creatures. Mollusks include octopi and squid, as well as snails and slugs, which are usually targeted by molluscicides.

Slugs and snails are well-known by farmers and gardeners as some of the most annoying and destructive pests. They can destroy the leaves and fruit of a large variety of plants, old and young, according to the University of California Integrated Pest Management Program.

"Snails and slugs feed on a variety of living plants and on decaying plant matter," UC IPM explained. "They chew irregular holes with smooth edges in leaves and flowers and can clip succulent plant parts. They also can chew fruit and young plant bark. Because they prefer succulent foliage or flowers, they primarily are pests of seedlings and herbaceous plants, but they also are serious pests of ripening fruits that are close to the ground such as strawberries, artichokes, and tomatoes. They also will feed on foliage and fruit of some trees; citrus are especially susceptible to damage."

Luckily, these pests are easily noticeable by the silvery trails of dried mucus left behind by their foot. When farmers and gardeners spot the telltale signs and damage left behind by snails and slugs, it's time to look to molluscicides for help.

As the U.S. Army Corps of Engineers explained, various types of molluscicides typically have similar modes of action. Molluscicides, sometimes referred to as bait, are touched or ingested by the slug or snail, then the chemicals manipulate the water balance of the animal. By affecting the amount of water in the mollusk's body, the molluscicides put the slugs or snails on the path to organ failure and inhibit healthy body functions.

There are two forms of chemical molluscicides: oxidizing and non-oxidizing. The non-oxidizing are among the most popular because they're more cost efficient. Although expensive in small quantities, non-oxidizing molluscicides are more effective in these small doses.

The Army Corps of Engineers outlined some of the most common non-oxidizing molluscicides including:

Quanternary and polyquaternary ammonium compounds

Aromatic hydrocarbons

Endothall as the mono salt

Metals like copper sulfate and their salts

Niclosamide.

Oxidizing compounds that can be used to combat snail and slug infestations include chlorine, chlorine dioxide, chloramines, ozone, bromine, hydrogen peroxide and potassium permanganate.

One of the most commonly used chemicals in molluscicides is metaldehyde. Used in Antimillace, Ariotox, Deadline, Halizan, Limatox, Namekil and Slug-Tox, this chemical has been popular since the late 1960s for attracting and killing slugs and snails.

Other uses for Molluscicides

Aside from slugs and snails in the garden or on a farm, mollusks can also harm commercial and industrial business in bodies of water, Marrone Bio innovations explained. Mollusks can clog pipes, damage equipment and costs businesses significant amounts of money in lost efficiency.

Often people turn to molluscicides to get rid of these pests and keep business profitable. However, there are a variety of environmental factors to consider to avoid wide-spread pollution in the waterway.

Acaricide

Acaricide is a kind of pesticide, which is specially used to prevent and control the harmful acarids in the arachnids. Such as Nisuo Lang, bromide mite, dicofol, dicofol, four mites, triazole tin, single formamidine, thiophene ketone, quinacridone, mite net, speed mites, etc. Most acaricide with contact action or systemic effect in different developmental stages of mites often show some selectivity, some acaricides on adult mites, young mites and eggs are effective, some can only kill mites and eggs invalid, some can only kill eggs, known as the ovicidal agent.Appropriate acaricides should be selected based on the type of mite and the period of control. Such as tetradifon has a strong contact effect on the mite eggs and larvae, but the effect of adult mites poor. A variety of organophosphorus insecticides, such as omethoate, monocrotophos, isocarbophos, also has acaricidal effect, and can be used in rotation with acaricides to prevent resistant mites or delay the production of drug resistance. After the experiment, Acaricides can be used together with pesticides and fungicides to combat a variety of pests and mites.

In 1944, the United States StaufferChemical Co. developed the one-chlorimicarb sulfone, which only acaricidal effect. Since then, the rapid development of special acaricide, new varieties continue to emerge. The main acaricides appearing in each period: in the late 1940s, there were dinocap, neotran, chlorfenson, aramite. In the 1950s, there were dimite, benzilan, three dicofol, three chlorocide sulfone. In the 1960s, there were binapacryl, dinobuton, propargite, morestan, bromopropylate, chlorfensulphide, Chlordimeform, cyhexatin. In the 1970s, there were citrazon, fenbutatin oxide, azocyclotin, amitraz, phenproxide. In the 1980s, there were acaridyl ether, thiophene, clofentezine, nisolang. In order to solve the problem of the rapid development of resistance to insecticides, people are constantly studying and developing a new type of insecticide, which can prevent and cure the insecticide resistant mites. In order to reduce the pollution of pesticides on the environment, it is a direction to develop high efficient acaricide. Nisorang's successful development is a good example. It is effective in the prevention and control of mites concentration of only 30-50 mg / L, valid for 60 to 70 days.

There are more varieties of acaricides, mainly organic chlorine, organic sulfur, nitrobenzene, organic tin, amidine, heterocyclic. Other types include sulfur, dinitrophenols, diphenyl alcohols, diphenoxymethanes, diphenylsulfonates, diphenylsulfonates, sulfinate esters, diphenyls Quinolines, sulfonamides, hydroxamates, amidines, benzimidazoles, phenylhydrazones, organotins, antibiotics, etc.

Acaricides are mostly low toxicity, relatively safe for humans and animals. The resistance rate of mites to the insecticide is relatively fast. There is usually no cross-resistance between different types of acaricides. In order to control drug-resistant mites and delay the development of resistance to mites, we should pay more attention to rotation use of different types of acaricide or mixed use. The existing acaricides have no absorption and conduction, spraying must be uniform and thoughtful. Some acaricides kill egg activity is high, do not kill into mites, and some are opposite, so the use of such two types of drugs often can improve the activity of killing mites. Special acaricides multi no killing effect.

Piscicide

The normal function of fish toxicants (piscicides) in fisheries management is to destroy unwanted fish. Attempts to destroy one species completely while protecting other species have generally been unsuccessful, although the use of TFM (trifluoromethyl nitrophenol) to control lampreys, Antimycin to control scalefish in rearing ponds for channel catfish, or Squoxin (1,1'-methylenedi-2-napthol) to destroy squawfish are examples of poisons which have proved reasonably successful in this respect.

Fewer investigations have been made into the use of chemicals to sample fish populations with the intention of obtaining information which may otherwise be difficult to acquire. Lennon et al. (1970) reviewed about 40 chemicals which have been used as toxicants for fish or other aquatic life, but only two such toxicants, Rotenone and Antimycin, seem to have been used for fish sampling. The first being probably the oldest and most commonly used of all piscicides, and the second one of the most recently-introduced types. This review attempts to summarize the main characteristics of both the piscicides and the techniques used for sampling.

Properties of Piscicides

All toxic chemicals used in fisheries management, even anaesthetics, are capable of causing death of fish if a sufficient concentration is used. No piscicide, however, is limited in its effects to those on fish, and many chemicals are lethal, at concentrations used on fish, to planktonic and benthic invertebrates, reptiles and amphibians, and occasionally toxic (but not necessarily lethal) to birds or mammals. The time of year at which the chemicals are used must be selected carefully. Some amphibians in Europe are in danger of extinction as the result of environmental pressures, while the destruction of fish which are a source of food for fish-eating birds could be serious at times when the young are being reared. Where the water being treated is destined for human consumption, both the potential toxicity and organoleptic properties must be considered.

Hall (1975) reported that sodium cyanide, cresol, copper sulphate and toxaphene have been used in waters of the Tennessee Valley Authority (TVA), U.S.A., in addition to rotenone, the most commonly-used fish toxicant. Antonioni and Baumann (1975) described the use of Antimycin A as a management and sampling tool. Of the others, sodium cyanide is unacceptable in water supply reservoirs in view of its mammalian toxicity, cresol is likely to cause tainting and toxaphene is undesirably persistent. Copper sulphate, at concentrations likely to kill fish, is very toxic to venegation. Meyer et al. (1976) reviewed the system introduced in the United States in 1972 for the registration of chemicals used in fisheries management. Rotenone and Antimycin were approved for use as piscicides, but fish contaminated by them are not acceptable for human consumption, primarily because no suitable methods are available for residue analysis.

Rotenone

Rotenone, the active constituent of derris root, is generally accepted as the chemical for fisheries management purposes with the fewest disadvantages in use. It is lethal, at concentrations used against fish, to many species of zooplankton and to some other invertebrates, but the effects on these elements of the fauna are not generally long-lasting. It is not dangerous to birds or mammals except at very high concentrations, but it, or the other chemicals present in commercial formulations, will taint water used for drinking purposes at concentration lethal to fish.

The stability or persistence of rotenone is dependent on pH, temperature, oxygen concentration and the presence of suspended matter, among other factors, but in practice, when used in lakes, the toxic effects usually last only for a period of a few days at summer temperatures. In winter, however, at near-zero temperatures the water of treated lakes can remain toxic for weeks (Holden, unpublished). Rotenone can be destroyed rapidly by treatment of the water with potassium permanganate, an effective technique which adds significantly to operational costs. An appropriate concentration of permanganate is 1 mg/l, but 5 mg/l may itself be toxic. Alternatively, activated carbon can be used in small but unpolluted streams (Engstrom-Heg, 1974). In dense weed beds, distribution of poison is difficult, and rotenone is rapidly destroyed.

As with all chemicals, sensitivity to rotenone varies appreciably between species, and between fry, juveniles and adults of one species, and is affected by temperature and certain chemical factors such as pH. Meadows (1973) found that to ensure a complete kill in a few hours it would be necessary to use 1 mg/l of a 5 percent rotenone formulation for common perch and at least 8 mg/l for crucian carp. The toxicity increased with temperature, but hardness had little influence. The concentration most commonly used of formulations containing 5 percent rotenone appears to be 0.5 or 1.0 mg/l (0.025–0.050 mg/l rotenone), which is capable of killing most fish within an hour or two at temperatures of 15°C or more, but at much higher temperatures death may result in a few minutes. At temperatures below 10°C, death may take several hours. As young fish are usually affected first, larger predatory fish may gorge on the smaller individuals before they themselves are affected.

Antimycin

Antonioni and Baumann (1975) described the merits of Antimycin as a fish poison, one advantage being that, unlike rotenone, fish do not avoid it. This could be an important advantage in some situations, when an escape route was available, or where natural springs or other areas of low concentration of toxin were present in which the fish could survive. The toxicity of Antimycin varies considerably with species, the LC_{100} value at 12°C being 0.8 microgrammes/1 for fish such as rainbow and brown trout, and yellow perch, but 120 microgrammes/1 for black bullheads. The latter concentration is of the same order as that usually used for rotenone. Marking (1975) found that the toxicity of Antimycin decreased gradually from pH 6.5 to pH 8.5, and abruptly above pH 8.5, which may limit its usefulness in some circumstances. Antonioni and Baumann found that Antimycin was unsuitable for use at sub-lethal concentrations as a sampling tool, as no behavioural abnormalities occurred in cases where ultimate death did not occur. They considered that detoxification with potassium permanganate was essential if apparently non-toxic concentrations were not to cause fish kills downstream to non-target species. It seems to have most promise in sampling rivers in conjunction with permanganate as a detoxifier. Antimycin has received less general approval than rotenone in many countries.

Practical Applications

The techniques employed in using rotenone or other piscicides for fish sampling can be sub-divided into those used in lakes or reservoirs, and those in rivers. One basic difference concerns the concentrations employed in the two situations. The time taken to kill fish in a lake, using a low concentration to minimize cost, may be a few hours, but in rivers it may be necessary to maintain a higher concentration for a much shorter period, usually no more than 30 minutes. On the other hand, the quantity of rotenone required for river sampling is usually less than that necessary for lakes, unless the areas sampled in the latter are small. In both types of situation, precautions must be taken to ensure that fish in the area poisoned do not escape from the area, and also that fish from outside the area cannot move into it during the sampling operation.

Rivers

Hocutt et al. (1970) used a block net in a river to retain all fish drifting down from the treated area above it, rotenone being added 68–91 metres upstream. Potassium permanganate was added immediately below the block net to prevent fish kills downstream. Although a complete sampling of the full width of the river could not be attempted, the technique was judged to be more successful than seine netting, as almost twice as many species of fish were collected. The 2.5 percent emulsified rotenone was added at the rate of 0.26 litres per cubic metre per second of water flow, over a period of 30 minutes, giving an average concentration of 0.15 mg/l of 2.5 percent rotenone over the period. This concentration could not be lethal, but presumably was sufficient to incapacitate the fish, which drifted into the block net.

Boccardy and Cooper (1963) used 2 litres of 5 percent emulsified rotenone per cubic metre per second of water inflow, a concentration over 30 minutes about fifteen times greater than Hocutt et al., who consider that the amount used by Boccardy and Cooper was excessive. it would seem that, in flowing waters, it may not be necessary to attempt to kill the fish in the period of the operation, but only to cause partial asphyxiation.

Standing Waters - Cove Sampling

In lakes, the most effective method of sampling appears to be to apply poison to bays which have been divided from the body of the lake by barrier nets, with a mesh sufficiently small to prevent the loss of fish above the minimum size to be counted. However, it must be recognized that species inhabiting littoral areas or bays are not necessarily representative of open or deep water. Hayne et al. (1968) found that cove samples gave species lists similar to the entire study areas examined, although less common species were not present in all coves. Size classes of abundant species were represented in both cove and lake samples, and standing stock estimates as biomass per unit area in coves were similar to those in the lake. Individual cove samples often gave under or over-estimates of numerical abundance of an individual species. Hall (1975), however, states that the information obtained closely reflects the situation in cove and shoreline areas only, although the littoral zone is considered to be the most productive in the lake, and is also where most fish and fishing activity occurs.

Calculation of the volume of water in the area selected must be accurately made, and the poison applied as evenly as possible. Hall (1975) described the method used in TVA reservoirs, in which the area immediately outside the block net is poisoned to reduce the movement of small fish into the area. The concentration of 5 percent emulsified rotenone used is 0.6–1.0 mg/l, the upper limit being more certain of producing a complete kill. For the usual area treated (about 0.4 ha) with an average depth of 5 m, the amount of 5 percent emulsified rotenone required is about 20 litres. The lakes described by Hall are warm-water lakes, with summer temperatures of 24–29°C, much higher than most European lakes. Davies et al. (1975) described an evaluation of the use of rotenone in sampling fish in a reservoir, in which the chemical was used on 29 occasions over five years in coves blocked off by nets. The fish were collected from the surface as quickly as possible and it was found that the technique gave a reasonable estimate of standing crop, and was useful in demonstrating the presence or absence of a species and the balance between groups of species. It was less useful, however, in determining year-class strength.

Cove sampling with rotenone was also used by Balon and Coche (1974) to provide samples of fish from the large man-made Lake Kariba in Africa. Sampling areas of 0.021–5.273 ha were selected as representative of the various parts of the lake, and fish density was found to be inversely related to the sampling area poisoned. Bazigos provides a detailed statistical analysis of the results obtained in an appendix to the report on Lake Kariba.

Standing Waters - Complete Poisoning

The technique of complete poisoning, to provide information on the total populations of fish in lakes or other water bodies which can be regarded as representative of similar lakes in the same area, has been used by several workers. HolĎik (1972) used a chemical (not identified) to determine the total biomass of fish and the species diversity in a number of shallow lakes, marshes, oxbows and water-holes in the alluvial plain of the Danube in Czechoslovakia. The water bodies were small, ranging from 8 m^2 up to 0.5 ha, and were treated during periods of low water level. All the fish were collected, and at the time of poisoning the stock was found to consist mainly of young-of-the-year specimens of phytophile species, the adults of which had migrated into the area during floods. Thus the populations determined could only be regarded as representative of a limited period of the year, as the spawning adults were apparently only temporarily resident. Loubens (1960) used a similar technique to study fish populations in floodplain waters of the Yaérés floodplains (Chad), and Kapetsky studied fish populations in the flood lakes of the Magadalena River using fish toxicants. Sumari (1975) used rotenone in 32 small Finnish lakes of 0.7–6.4 ha, applying sufficient quantities to kill the entire population in each lake. These lakes were selected as being representative of the much larger number in the area. SCUBA divers collected the fish from 100 m^2 areas marked out on the bottom of each lake, the depth in most cases being less than 5 m. A greater relative proportion of small fish and bottom-feeding species were found on the bottom. In the lakes studied, an average of 26 percent of the biomass of ruff, 50 percent of burbot, 81 percent of pike, 74 percent of perch and 66 percent of roach were collected from the surface.

Dicofol

Mites are tiny, spider-like creatures that thrive in the same rich, moist soils that many crops do. Because they cohabitate with plants in some environments, they're also a major pest to farmers. They can infest and damage a number of plant varieties, hurting farmers' yields and costing them money. To fight these very small pests, farmers use miticides, like dicofol.

Understand Dicofol

Dicofol is an organochlorine pesticide that's effective against mites, but controversial because of its environmental impact. Written as $C_{14}H_9Cl_5O$ or 2,2,2-trichloro-1,1-bis(4-chlorophenyl)ethano, this acaricide is used on a variety of crops including cotton, apples, citrus fruits, mint, strawberries, beans, peppers, tomatoes, walnuts, stone fruits, cucumbers, ornamentals, and other fruits and vegetables.

Among the various mites that dicofol is effective against, it is often used on the red spider mite.

"Dicofol is used to kill crop-feeding mite pests such as the red spider mite. It is a contact poison which kills the pest after being ingested and picked up from the surface of the crop. In many countries, dicofol is also used in combination with other pesticides such as the organophosphates parathion-methyl, and dimethoate," according to the pesticide information site.

There are a number of commercial varieties that feature this miticide such as Acarin, Cekudifol, Decofol, Dicaron, Dicomite, Difol, Hilfol, Kelthane, and Mitigan. In its purest form, dicofol is a white crystal, however in most applications it appears as an amber-colored, viscous liquid that smells like hay.

According to Oregon State University, dicofol was outlawed for a brief period in the 1980s because of the way it was made and how much DDT was involved. Initially, the creation of dicofol would create a substantial amount of DDT, which seriously impacted the environment. Dicofol is chemically similar to DTT – C14H9Cl5 – which is lacking the oxygen component of its miticide relative.

Impurities

Manufacturing-use dicofol products contain a number of DDT analogs as manufacturing impurities. These include the o,p' and p,p' isomers of DDT, DDE, DDD, and a substance called extra-chlorine DDT or Cl-DDT.

Use and Formulations

Foliar spray on agricultural crops and ornamentals, and in or around agricultural and domestic buildings for mite control. It is formulated as emulsifiable concentrates, wettable powders, dusts, ready-to-use liquids, and aerosol sprays. In many countries, dicofol is also used in combination with other pesticides such as the organophosphates, methyl parathion, and dimethoate.

Producers

Dicofol first appeared in the scientific literature in 1956, and was introduced onto the market by the US-based multinational company Rohm & Haas in 1957. Other current manufacturers include Hindustan Insecticides Limited (India), Lainco (Spain), and ADAMA Agricultural Solutions (Formerly Makhteshim-Agan) (Israel). It is sold under a number of trade names, including *Hilfol*, *Kelthane* and *Acarin*.

In 1986, the US Environmental Protection Agency (EPA) temporarily canceled the use of dicofol because relatively high levels of DDT contamination were ending up in the final product. Modern processes can produce technical grade dicofol that contains less than 0.1% DDT.

Estimated usage as a Pesticide

The Pesticide Survey, USA 1987 through 1996, reports that the total annual domestic agricultural usage of dicofol averaged about 860,000 pounds active ingredient (a.i.) for about 720,000 acres (2,900 km^2) treated. Most of the area is treated with 2 pounds a.i. or less per application, and the average acre is treated with about 1.2 pounds a.i. per year (1.3 kg/(ha·yr)). Fruits tend to have the highest application rates. The largest markets for dicofol in terms of total pounds active ingredient are cotton (over 50%) and citrus (almost 30%). Although only about 4% of the cotton acres grown are treated with dicofol, over 60% of all crop acres treated with dicofol are cotton acres. The remaining usage is primarily on other fruits and vegetables. Most of the US usage is in California and Florida.

Effects

The California Department of Food and Agriculture has one of the world›s most extensive incident reporting systems. Between 1982 and 1992, 38 incidents involving dicofol alone were reported: systemic 19 (50%); skin 10 (26%); eye 8 (21%); and eye/skin 1 (3%). The number of incidents per 1,000 applications for all illnesses ranged from 0.11 to 0.21.

The US National Pesticides Telecommunications Network database collected reports from 1984 to 1991 showing 91 human, 9 animal and 31 other poisoning incidents for a total of 131 incidents involving dicofol from 571 phone calls made to the hotline.

An assessment of dicofol by the UK Pesticides Safety Directorate in 1996 found that residues in apples, pears, blackcurrants and strawberries were higher than expected.

There is no established US maximum contaminant level (MCL) or health advisory levels for residues of dicofol in drinking water. In the European Union, the maximum level is the same for all active ingredients 0.1 mg/l.

In 1990, the use of dicofol was suspended in Sweden for environmental reasons. In Switzerland its use is permitted for research purposes only. Throughout the European Union dicofol containing more than lg/kg (0.1%) of DDT or DDT related compounds cannot be used.

The 1998 US EPA review of dicofol recommended a number of changes in order to protect the environment and wildlife. Dicofol applications are limited to no more than one per year. In the UK, the maximum number of treatments permitted is two per year for apples and hops, and two per crop for strawberries, protected crops and tomatoes.

In 1980, an accident at the US Tower Chemical Company led to a release of dicofol into Lake Apopka in Florida. Ten years later Dr Guillette of Florida University linked this incident to a subsequent decline in the fertility of alligators in the lake. The US EPA is still not clear whether dicofol is involved in the reproductive failure of the alligator population following the accidental spill.

Toxicity

It is classified by the World Health Organisation as a Class II, 'moderately hazardous' pesticide.

The acute oral LD_{50} for dicofol is 587 mg/kg for rats.

Dicofol is a nerve poison. The exact mode of action is not known, although in mammals it causes hyperstimulation of nerve transmission along nerve axons (cells). This effect is thought to be related to the inhibition of certain enzymes in the central nervous system.

Symptoms of ingestion and/or respiratory exposure include nausea, dizziness, weakness and vomiting; dermal exposure may cause skin irritation or a rash; and eye contact may cause conjunctivitis. Poisoning may affect the liver, kidneys or the central nervous system. Very severe cases may result in convulsions, coma, or death from respiratory failure.

Dicofol can be stored in fatty tissue. Intense activity or starvation may mobilize the chemical, resulting in the reappearance of toxic symptoms long after actual exposure.

Chronic Effects

Tests on laboratory animals show that the primary effects after long term exposure to dicofol include increases in liver weight and enzyme induction in the rat, mouse and dog.

There are also effects relating to altered adrenocorticoid metabolism (part of the hormonal system). In the rat hormonal changes were accompanied by the histological observation of vacuolation (empty cavities) of the cells of the adrenal cortex.

Carcinogenecity

The US EPA has classified dicofol as a Group C, possible human carcinogen. There is limited evidence that it may cause cancer in laboratory animals, but there is no evidence that it causes cancer in humans. This classification was based on animal test data that showed an increase in the incidence of liver adenomas (benign tumour) and combined liver adenomas and carcinomas in male mice.

Reproductive Effects

Reproductive effects in rat offspring have been observed only at doses high enough to also cause toxic effects on the livers, ovaries, and feeding behavior of the parents. Rats fed diets containing dicofol through two generations exhibited adverse effects on the survival and/or growth of newborns at 6.25 and 12.5 mg/kg/day

- Teratogenic effects: No teratogenic effects are observed when rats were given up to 25 mg/kg/day on days 6 through 15 of pregnancy
- Mutagenic effects: Laboratory tests have shown that dicofol is not mutagenic
- Endocrine disruption: Evidence for dicofol to cause endocrine disruption is suggestive, but not definitive

A 2007 study by the California Department of Public Health found that women in the first eight weeks of pregnancy who live near farm fields sprayed with dicofol and the related organochloride pesticide endosulfan are several times more likely to give birth to children with autism. These results are highly preliminary due to the small number of women and children involved and lack of evidence from other studies.

Metabolism

Dicofol is converted in rats to the metabolites 4,4›-dichlorobenzophenone and 4,4›-dichlorodicofol.

Studies of the metabolism of dicofol in rats, mice, and rabbits have shown that ingested dicofol is rapidly absorbed, distributed primarily to fat, and readily eliminated in feces. When mice were given a single oral dose of 25 mg/kg dicofol, approximately 60% of the dose was eliminated within 96 hours, 20% in the urine, and 40% in the feces. Concentrations in body tissues peaked between 24 and 48 hours following dosing, with 10% of the dose found in fat, followed by the liver and other tissues. Levels in tissues other than fat declined sharply after the peak.

Environmental Impact

Dicofol was brought back after a new manufacturing process showed that the chemical had less than 0.1 percent DDT. It is now approved for many fruits, vegetables and field crops, but the U.S. Environmental Protection Agency asked manufacturers to stop producing the chemical for lawn uses and they complied.

The EPA suggested people stop using dicofol on lawns because of the potential environmental repercussions. In 2006, the EPA created an amended Reregistration Eligibility Decision for the chemical which has since drastically cut down on its use. In 2011, the EPA requested voluntary cancelation of all dicofol uses in the U.S. by late 2016.

Dicofol is the last organochlorine pesticide to go through these processes in the U.S., and its likely all due to the environmental concerns. As OSU explained, there are several impacts the chemical can have on different environments and organisms.

- Birds – Dicofol is a dangerous substance to most birds. It is tied to eggshell thinning and reduced fertility, OSU explained. Quail, pheasant, dove and screech owl are among the affected birds.
- Waterlife – This chemical is significantly more toxic to aquatic life. Small animals like shrimp are specifically at risk as well as algae, fish and invertebrates.
- Bees – Dicofol is not toxic to bees and there are no known negative side effects.
- Water – This pesticide absorbs into the sentiment in most water, OSU explained, but it also degrades in UV light. It is insoluble in water and unlikely

to leech through the soil in great depths. A more likely route to water pollution is through erosion or improper application.

- Land and vegetation – Dicofol has a 60-day half-life in soil and is degradable in UV light. It is found in soil but only on the surface. The chemical can remain in plants for up to two years, according to OSU.
- Humans – Long- or short-term intake of residual dicofol is unlikely to present any danger to human health, according to the EPA.

References

- Miller GT (2002). Living in the Environment (12th ed.). Belmont: Wadsworth/Thomson Learning. ISBN 978-0-534-37697-0
- Gilden RC, Huffling K, Sattler B (January 2010). "Pesticides and health risks". Journal of Obstetric, Gynecologic, and Neonatal Nursing. 39 (1): 103–10. doi:10.1111/j.1552-6909.2009.01092.x. PMID 20409108
- Cornell University. Toxicity of pesticides. Pesticide fact sheets and tutorial, module 4. Pesticide Safety Education Program. Retrieved on 2007-10-10
- Gilden RC, Huffling K, Sattler B (January 2010). "Pesticides and health risks". Journal of Obstetric, Gynecologic, and Neonatal Nursing. 39 (1): 103–10. doi:10.1111/j.1552-6909.2009.01092.x. PMID 20409108
- Yamamoto I (1999). "Nicotine to Nicotinoids: 1962 to 1997". In Yamamoto I, Casida J. Nicotinoid Insecticides and the Nicotinic Acetylcholine Receptor. Tokyo: Springer-Verlag. pp. 3–27. ISBN 978-4-431-70213-9
- Lamberth C, Jeanmart S, Luksch T, Plant A (August 2013). "Current challenges and trends in the discovery of agrochemicals". Science. 341 (6147): 742–6. doi:10.1126/science.1237227. PMID 23950530
- Goldman LR (2007). "Managing pesticide chronic health risks: U.S. policies". Journal of Agromedicine. 12 (1): 67–75. doi:10.1300/J096v12n02_08. PMID 18032337
- Encyclopedia of environment and society. Robbins, Paul, 1967-, Sage Publications. Thousand Oaks. p. 862. ISBN 9781452265582. OCLC 228071686
- Kolberg, Robert L.; Lori J. Wiles (2002). "Effect of Steam Application on Cropland Weeds1". Weed Technology. 16: 43–49. doi:10.1614/0890-037X(2002)016[0043:EOSAOC]2.0.CO;2
- IMAZAMOX | Pacific Northwest Weed Management HandbookArchived 2012-06-25 at the Wayback Machine.. Pnwhandbooks.org. Retrieved 2013-03-05
- Beckie, H. J.; Harker, L. M.; Hall, S. I.; et al. (2006). "A decade of herbicide-resistant crops in Canada". Canadian Journal of Plant Science. 86 (4): 1243–1264. doi:10.4141/P05-193

Chapter 5

Insecticides and their Role in Pest Control and Management

Insecticides are chemicals that are used to kill insects in agriculture and in industry. They are generally distinguished into three types, such as natural, organic and inorganic insecticides. This chapter has been carefully written to provide an easy understanding of the varied types of insecticides and their role in pest control and management.

Insecticide

Insecticide, any toxic substance that is used to kill insects. Such substances are used primarily to control pests that infest cultivated plants or to eliminate disease-carrying insects in specific areas.

Insecticides can be classified in any of several ways, on the basis of their chemistry, their toxicological action, or their mode of penetration.They are classified according to whether they take effect upon ingestion (stomach poisons), inhalation (fumigants), or upon penetration of the body covering (contact poisons). Most synthetic organic insecticides penetrate by all three of these pathways, however, and hence are better distinguished from each other by their basic chemistry. Besides the synthetic organics, some organic compounds occurring naturally in plants are useful insecticides, as are some inorganic compounds. Most insecticides are sprayed or dusted onto plants and other surfaces traversed or fed upon by insects.

Modes of Penetration

Stomach poisons are toxic only if ingested through the mouth and are most useful against those insects that have biting or chewing mouth parts, such as caterpillars, beetles, and grasshoppers. The chief stomach poisons are the arsenicals—e.g., Paris green (copper acetoarsenite), lead arsenate, and calcium arsenate; and the fluorine compounds, among them sodium fluoride and cryolite. They are applied as sprays or dusts onto the leaves and stems of plants eaten by the target insects. Stomach poisons have gradually been replaced by synthetic organic insecticides, which are less dangerous to humans and other mammals.

Contact poisons penetrate the skin of the pest and are used against those arthropods, such as aphids, that pierce the surface of a plant and suck out the juices. The contact insecticides can be divided into two main groups: naturally occurring compounds and synthetic organic ones. The naturally occurring contact insecticides include nicotine, developed from tobacco; pyrethrum, obtained from flowers of *Chrysanthemum cinerariaefolium;* rotenone, from the roots of *Derris* species and related plants; and oils, from petroleum. Though these compounds were originally derived mainly from plant extracts, the toxic agents of some of them (e.g., pyrethrins) have been synthesized. Natural insecticides are usually short-lived on plants and cannot provide protection against prolonged invasions. Except for pyrethrum, they have largely been replaced by newer synthetic organic insecticides.

Fumigants are toxic compounds that enter the respiratory system of the insect through its spiracles, or breathing openings. They include such chemicals as hydrogen cyanide, naphthalene, nicotine, and methyl bromide and are used mainly for killing insect pests of stored products or for fumigating nursery stock.

Type of Activity

Systemic insecticides become incorporated and distributed systemically throughout the whole plant. When insects feed on the plant, they ingest the insecticide. Systemic insecticides produced by transgenicplants are called plant-incorporated protectants (PIPs). For instance, a gene that codes for a specific *Bacillus thuringiensis* biocidal protein was introduced into corn (maize) and other species. The plant manufactures the protein, which kills the insect when consumed.

Contact insecticides are toxic to insects upon direct contact. These can be inorganic insecticides, which are metals and include arsenates, copper and fluorine compounds, which are less commonly used, and the commonly used sulfur. Contact insecticides can be organic insecticides, i.e. organic chemical compounds, synthetically produced, and comprising the largest numbers of pesticides used today. Or they can be natural compounds like pyrethrum, neem oil etc. Contact insecticides usually have no residual activity.

Efficacy can be related to the quality of pesticide application, with small droplets, such as aerosols often improving performance.

Biological Pesticides

Many organic compounds are produced by plants for the purpose of defending the host plant from predation. A trivial case is tree rosin, which is a natural insecticide. Specifically, the production of oleoresin by conifer species is a component of the defense response against insect attack and fungal pathogen infection. Many fragrances, e.g. oil of wintergreen, are in fact antifeedants.

Four extracts of plants are in commercial use: pyrethrum, rotenone, neem oil, and various essential oils.

Other Biological Approaches

Plant-incorporated Protectants

Transgenic crops that act as insecticides began in 1996 with a genetically modified potato that produced the Cry protein, derived from the bacterium Bacillus thuringiensis, which is toxic to beetle larvae such as the Colorado potato beetle. The techniquc has been expanded to include the use of RNA interference RNAi that fatally silences crucial insect genes. RNAi likely evolved as a defense against viruses. Midgut cells in many larvae take up the molecules and help spread the signal. The technology can target only insects that have the silenced sequence, as was demonstrated when a particular RNAi affected only one of four fruit fly species. The technique is expected to replace many other insecticides, which are losing effectiveness due to the spread of pesticide resistance.

Enzymes

Many plants exude substances to repel insects. Premier examples are substances activated by the enzyme myrosinase. This enzyme converts glucosinolates to various compounds that are toxic to herbivorousinsects. One product of this enzyme is allyl isothiocyanate, the pungent ingredient in horseradish sauces.

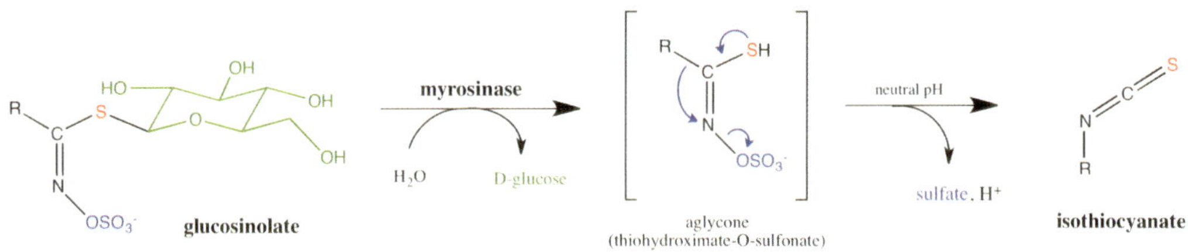

Biosynthesis of antifeedants by the action of myrosinase.

The myrosinase is released only upon crushing the flesh of horseradish. Since allyl isothiocyanate is harmful to the plant as well as the insect, it is stored in the harmless form of the glucosinolate, separate from the myrosinase enzyme.

Bacterial

Bacillus thuringiensis is a bacterial disease that affects Lepidopterans and some other insects. Toxins produced by strains of this bacterium are used as a larvicide against caterpillars, beetles, and mosquitoes. Toxins from *Saccharopolyspora spinosa* are isolated from fermentations and sold as Spinosad. Because these toxins have little effect on other organisms, they are considered more environmentally friendly than synthetic pesticides. The toxin from *B. thuringiensis* (Bt toxin) has been incorporated directly into plants through the use of genetic engineering. Other biological insecticides include products based on entomopathogenic fungi (e.g., *Beauveria bassiana, Metarhizium anisopliae*), nematodes (e.g., *Steinernema feltiae*) and viruses (e.g., *Cydia pomonella* granulovirus).

Synthetic Insecticide and Natural Insecticides

A major emphasis of organic chemistry is the development of chemical tools to enhance agricultural productivity. Insecticides represent a major area of emphasis. Many of the major insecticides are inspired by biological analogues. Many others are completely alien to nature.

Organochlorides

The best known organochloride, DDT, was created by Swiss scientist Paul Müller. For this discovery, he was awarded the 1948 Nobel Prize for Physiology or Medicine. DDT was introduced in 1944. It functions by opening sodium channels in the insect›s nerve cells. The contemporaneous rise of the chemical industry facilitated large-scale production of DDT and related chlorinated hydrocarbons.

Organophosphates and Carbamates

Organophosphates are another large class of contact insecticides. These also target the insect›s nervous system. Organophosphates interfere with the enzymes acetylcholinesterase and other cholinesterases, disrupting nerve impulses and killing or disabling the insect. Organophosphate insecticides and chemical warfare nerve agents (such as sarin, tabun, soman, and VX) work in the same way. Organophosphates have a cumulative toxic effect to wildlife, so multiple exposures to the chemicals amplifies the toxicity. In the US, organophosphate use declined with the rise of substitutes.

Carbamate insecticides have similar mechanisms to organophosphates, but have a much shorter duration of action and are somewhat less toxic.

Pyrethroids

Pyrethroid pesticides mimic the insecticidal activity of the natural compound pyrethrum, the biopesticide found in pyrethrins. These compounds are nonpersistent sodium channel modulators and are less toxic than organophosphates and carbamates. Compounds in this group are often applied against household pests.

Neonicotinoids

Neonicotinoids are synthetic analogues of the natural insecticide nicotine (with much lower acute mammalian toxicity and greater field persistence). These chemicals are acetylcholine receptor agonists. They are broad-spectrum systemic insecticides, with rapid action (minutes-hours). They are applied as sprays, drenches, seed and soil treatments. Treated insects exhibit leg tremors, rapid wing motion, styletwithdrawal (aphids), disoriented movement, paralysis and death. Imidacloprid may be the most common. It has recently come under scrutiny for allegedly pernicious effects on honeybees and its potential to increase the susceptibility of rice to planthopper attacks.

Ryanoids

Ryanoids are synthetic analogues with the same mode of action as ryanodine, a naturally occurring insecticide extracted from *Ryania speciosa* (Flacourtiaceae). They bind to calcium channels in cardiac and skeletal muscle, blocking nerve transmission. The first insecticide from this class to be registered was Rynaxypyr, generic name chlorantraniliprole.

Insect Growth Regulators

Insect growth regulator (IGR) is a term coined to include insect hormone mimics and an earlier class of chemicals, the benzoylphenyl ureas, which inhibit chitin (exoskeleton) biosynthesis in insectsDiflubenzuron is a member of the latter class, used primarily to control caterpillars that are pests. The most successful insecticides in this class are the juvenoids (juvenile hormone analogues). Of these, methoprene is most widely used. It has no observable acute toxicity in rats and is approved by World Health Organization (WHO) for use in drinking water cisterns to combat malaria. Most of its uses are to combat insects where the adult is the pest, including mosquitoes, several fly species, and fleas. Two very similar products, hydroprene and kinoprene, are used for controlling species such as cockroaches and white flies. Methoprene was registered with the EPA in 1975. Virtually no reports of resistance have been filed. A more recent type of IGR is the ecdysone agonist tebufenozide (MIMIC), which is used in forestryand other applications for control of caterpillars, which are far more sensitive to its hormonal effects than other insect orders.

Examples

Organochlorides

Organophosphates

Carbamates

Pyrethroids

Neonicotinoids

Ryanoids

Insect growth regulators

Derived from plants or microbes

Some insecticides, such as insecticidal soaps, work through non-toxic methods, but others contain chemicals that are harmful to other forms of life. Insecticides are also ineffective when used excessively because pests build up resistance. Non-chemical control methods don't have these disadvantages.

Non-Target Organisms

Insecticides can affect many more organisms than the ones they're intended to kill. Insecticides are not without risks to the humans who use them. And when insecticides enter water sources through runoff, drift, leaching or accidental release, they kill fish and other aquatic wildlife. Birds die from drinking contaminated water and eating affected insects, as well as from accidental exposure to sprays. The long-term effects of insecticides on non-target organisms are unknown. Some remain in the bodies of animals that eat them and travel up the food chain, becoming more concentrated until finally affecting top predators. Some insecticides, including DDT, were largely banned in the United States because they were found to affect the reproductive abilities of predatory birds.

Beneficial Insects

Insecticides can kill insects that help gardeners and farmers. Honeybees are essential for crop pollination as well as making honey, yet they and other pollinators are killed by widely-used insecticides. Insecticides also kill ladybugs, predatory wasps, lacewings, hoverflies, ground beetles, spiders and other insects that eat pests. Gardeners sometimes make infestations worse by killing a pest's natural predators. The wider roles insects and microorganisms that feed on pests play in agriculture and the environment are probably underestimated, states the Institute of Food and Agricultural Sciences, University of Florida.

Resistance

Repeated exposure to insecticides builds up resistance in insects, until finally the insecticide has little or no effect. Every insect population produces some individuals with a genetic makeup that allows them to withstand the insecticide's effects. These individuals survive the spraying and pass on their protective genetic makeup to their offspring. Over time, the population of resistant individuals increases until it forms the majority. Insects reproduce quickly, often as frequently as a new generation every three or four weeks, so resistance can build up rapidly. Frequent insecticide applications make the problem worse.

Minimizing Effects

Using selective, non-persistent insecticides and taking precautions when spraying help minimize the disadvantages of insecticides. Selective insecticides kill only a narrow range of organisms, and non-persistent insecticides break down quickly in the environment. Use ready-mixed products to avoid having to contaminate other containers with pesticide, and spray only where the pest attack is most severe. Don't spray on windy days or when rain is forecast, and spray in the early morning to have the least impact on bees. Non-chemical pest control methods include optimizing growing conditions, growing resistant varieties, using insect traps or barriers, hosing plants down and growing plants that attract beneficial insects.

Inorganic Insecticides

Inorganic Insecticides are insecticides that are either man-made or based on synthetic formulas like carbamates or pyrethroids. These are made from heavy metals and arsenic compounds such as boric acid and silica gel. Inorganic types of insecticides are highly effective but use them carefully and read the applying conditions properly. There are several types of inorganic insecticides.

Systematic Insecticides are put in the soil and allowed to be absorbed by plant roots. These insecticides move up to the leaves, branches, fruits and twigs of the plant and protect it from being chewed upon by insects. Bactericides and larvicides are also a form of systematic insecticides that help in eradicating bacteria and larvae that destroys plants.

Contact Insecticides are the common sprays that can be used at home to eradicate insects and mosquitoes. Some of these products are also used to disinfect the home and environment and deodorise the place. These insecticides are available in the market in the form of sprays, coils, liquid vaporisers and repellents.

Ingested Insecticides are used to kill rats, rodents and cockroaches. Rodenticide is also a form of ingested insecticide. It is sort of a poison that is effective in killing rats. It is used as poisonous bait and fed to rats to get rid of them.

Herbicides are yet another kind of pesticides that are used to kill infected plants. It is specifically used to destroy parts of the plant that the insects feed on or the sections of the plant where infection has spread. Fungicides are also from the same family of herbicides; they are used to kill weeds and fungi that are responsible for destroying the agricultural yield.

Paris Green

Paris green, also called Schweinfurt green, an extremely poisonous, bright green powder that was formerly used extensively as a pigment (e.g., in wallpaper) and that is sometimes used as an insecticide or to kill plant fungi it must be used with great caution because of its poisonous nature. Chemically it is a copper acetoarsenite that may be prepared from arsenic trioxide and copper acetate.

Properties

Paris green is a blue-green solid. It is extremely toxic. It has a very bright blue-green color. It darkens in air.

Preparation

It is made by reacting copper(II) acetate with arsenic trioxide and water.

Uses

Paris green was first used to kill rats in sewers in Paris. This is where it got its name from. It was used as an insecticide. It was dropped from airplanes to stop malaria in Italy. It was used as a bright green pigment. It was painted on ships to prevent barnacles from growing. In many places, Paris green is banned because it is so toxic. It is only used in a very few places.

Safety

Paris green is extremely toxic. It used to be common and so was a common poison.

Boric Acid

Boric acid, also called orthoboric acid, boracic acid, hydrogen borate and acidum boricum, is a low-toxicity mineral with insecticidal, fungicidal, and herbicidal properties. It is the "*secret ingredient*" in many of the commercial products used for insect control. It is also a weak acid of boron, often used as a flame retardant, antiseptic, or precursor to other chemical compounds.

Its chemical formula is H_3BO_3, and exists in the form of a white powder or colourless crystals that dissolves in warm water. It does not evaporate or volatilise into the air.

Boron, the main element of boric acid, doesn't exist in nature by itself. It must be combined with other elements to create compounds. Boron is an essential micro nutrient for plants and humans. We get 1-3 milligrams of boron every day from the food we eat. Grains, nuts, fruits and vegetables are good sources of boron.

Boric acid is known as a desiccant which means it induces or sustains a state of dryness in its local vicinity. It acts as a insect killer by removing moisture from the body causing severe dehydration and eventually death. It maintains its potency for a long period of time and is highly effective in extremely small amounts.

While it is not as fast acting as synthetic pesticides, it does not pose the considerable health concerns as the synthetic pesticides and remains highly effective over a long period of time.

It is odourless and, unlike other insecticides, is non-repellent. That means that insects will continue to return to the treated areas.

Effectiveness

Boric acid is excellent for killing insects. It works as a desiccant to dehydrate the insects, eventually causing death. It also attaches to the insects' feet as they walk through it allowing it to be taken back to the nest. It is neither an attractant nor a repellent so the insects will keep coming back to treated areas.

Within 2-3 weeks you should start to see some results and it can take up to a month to spread through an entire colony.

Boron is an essential nutrient for growth and development of healthy plants. Boron compounds are used in small concentrations as micronutrients in fertilizers. When used in large concentrations they function as herbicides, algaecides and other pesticides.

Functions of boron in plants:

- Essential for maintaining a balance between sugar and starch and helps in the translocation of sugar and carbohydrates
- Important in pollination and seed production
- Necessary for normal cell division, nitrogen metabolism and protein formation

Boron is an essential element for plants' development, growth, crop yielding and seed development. It helps with the transfer of water and nutrition in plants. Though plants' boron requirements are very low in quantity, their growth and crop yields are severely affected when soil is boron deficient.

Boric acid is commonly used in the agriculture industry as a source of boron for liquid fertilizers. These liquid boron fertilizers are typically about 10% boron, and are used for soil and foliar applications.

Borates in solid form can be applied directly onto soil, where in humid conditions they will readily dissolve then be taken up by the plants' roots. Typical products used for this application are granular borax pentahydrate (Etibor 48) and granular ulexite. These products can be applied directly, but are more commonly used as part of a nutrient blend containing other fertilizer components. The amount of boron fertilizer to be applied onto soil varies according to plant type, method of application, amount of rain, and the soil's lime and organic material contents.

Boron products can also be applied by spraying onto plant leaves. This benefits some plants whose intake is better through their leaves or in other cases where spraying is a better way of fertilizing. Disodium octaborate tetrahydrate (Etidot 67), which is specially made for agricultural applications, is the preferred boron product for these foliar sprays because it has much better solubility compared to the conventional boron products like borax decahydrate and borax pentahydrate, and a higher boron content than boric acid.

Other products used in fertilizer include borax decahydrate and borax pentahydrate. Because of their good solubility, sodium borates can be applied directly onto soil or sprayed onto plants. Ulexite, which is a naturally occurring calcium borate, is favored in sandy soils because of its low solubility. This helps it to remain in soil longer than sodium borates.

Boron-based herbicides are produced from borax and boric acid. They are generally mixed with sodium chloride or other chemicals.

Aluminium Phosphide

Aluminum Phosphide is a useful inorganic compound used as a pesticide. It is mainly used as a fumigant for pest control and as a source of Phosphine gas (also used for fumigation). This inorganic compound can also be used as a wide band gap or energy gap semiconductor.

It is a reducing agent, which indicates that it acts as an electron donor in a chemical reaction.

The usefulness of this compound as a source of Phosphine gas or Hydrogen Phosphide (PH3) was first discovered by a German company named Degesch. In 1958, the United States registered it as a pesticide to the Hollywood Termite Control Company, Inc.

Aluminum Phosphide Applications

It can be used as:

- Fumigant
- Insecticide
- Rodenticide
- Semiconductor

Uses of Aluminum Phosphide as Pesticide in Fumigation

Fumigation is the process of pest control for grains, soil, buildings and other different stored goods. In this case, the AlP is left to react with water and the resultant Phosphine gas is then used as the fumigant.

In case of fumigation of grains stored in silos, bins or any other similar container, 2- 4 Aluminum Phosphide tablets should be applied per cubic meter. If the grain is stored in a flat storage area, 3- 5 tablets per cubic meter will be adequate. In case of packed food or other stored goods, it will be sufficient to apply 1 to 1.5 tablets per cubic meter.

This inorganic chemical is used to control both the insects and rodents. Following are some information about its uses in the field of pest control:

Insecticide

Insecticide refers to a specific type of pesticide that is used only against insects. AlP is very popular as an insecticide as it is highly effective and very easy to use. The AlP insecticides are usually non-flammable, but it reacts with air moisture and creates PH3 gas which is very toxic and flammable. Due to this reason, these insecticides also contain suitable stabilizers, usually Ammonium Carbamate, to control the development of the poisonous gas.

Rodenticide

It indicates another type of pesticide that is specifically used against rodents. It is widely used in the US and is among the most poisonous pesticides. Aluminum Phosphide rodenticide can be very effective in driving away small mammals like rats, rabbits and moles. The rodenticide pellets are mixed with their food. Once they ingest it, the chemical reacts with the acids in the digestive system of the animal and produces PH3. As a result, the animal gets poisoned.

The use of this material as a fumigant has caused accidental deaths in many countries like Saudi Arabia and the US. The Iranian Forensic Medicine Organization has started a campaign to stop the use of this chemical as a pesticide.

Organochloride Insecticides

Organochloride pesticides are very effective at eliminating pests, especially insects. But many of these chemical products are viewed negatively by environmental activists and consumers because of one well-known and now banned organochloride pesticide: dichloro diphenyl trichoroethane, better known as DDT.

Organochloride pesticides refer to chemicals with carbon, chlorine and hydrogen components. As the U.S. Fish and Wildlife Service explained, chlorine-carbon bonds are particularly strong, which keeps these chemicals from breaking down quickly or being soluble in water. The chemical is also attracted to fat and is stored in the fatty tissue of animals that consume it. The durability of organochloride pesticides' chemical makeup is one of the reasons it is so effective as an insecticide as well as potentially harmful – it can protect crops for a long time, but also stay in an animal's system.

Along with DDT, other organochloride pesticides, such as aldrin, dieldrin, heptachlor, mirex, chlordecone and chlordane, have been banned by the U.S. Environmental Protection Agency. Europe has similarly banned many organochloride pesticides, but in both of these regions organochloride chemicals are still the active ingredients in a number of home, garden and environmental pest control products, according to the EPA. Organochloride pesticides are also enormously popular in developing countries across the globe for agricultural uses.

Testing for Organochloride Pesticides

Whether examining farm land to see if it is still filled with years-old organochloride pesticides or inspecting water for organochlorines, testing is the best way to know if you have these chemicals near you. EPA methods 8250A and 8270B can be used to test for these chemicals. 8250A can test waste, soil and water, while 8270B uses gas chromatography/mass spectrometry (GC/MS).

Although organochloride pesticides are most famous for damaging some birds' abilities to lay healthy eggs, these chemicals have been known to negatively affect humans who consume or breathe in pesticides. Accidental inhalation or consumption of a contaminated fish or animal tissue are the most likely ways to ingest organochloride pesticides. To confirm that someone is poisoned by a organochloride pesticide, blood or urine is typically sent to a university or government facility that uses GC/MS to test for the chemical compounds.

Warning signs of organochloride pesticide toxicity include seizures, hallucinations, coughing, skin rash, vomiting, abdominal pain, headaches, confusion and possibly respiratory failure, according to Matthew Wong, M.D., M.P.H., Beth Israel Deaconess Medical Center, on Medscape. Although there are bans in the U.S. and Europe on many of these pesticides, their use in other parts of the world and storage in parts of the U.S. and Europe create situations where poisonings are still possible.

Organochloride pesticides encompass a large number of durable chemicals that are both effective and carry a fair amount of risk across the globe.

Physical and Chemical Properties

Chlorination modifies the physical properties of hydrocarbons in several ways. The compounds are typically denser than water due to the higher atomic weight of chlorine versus hydrogen. Aliphatic organochlorides are alkylating agents because chloride is a leaving group.

Natural Occurrence

Many organochlorine compounds have been isolated from natural sources ranging from bacteria to humans. Chlorinated organic compounds are found in nearly every class of biomolecules including alkaloids, terpenes, amino acids, flavonoids, steroids, and fatty acids. Organochlorides, including dioxins, are produced in the high temperature environment of forest fires, and dioxins have been found in the preserved ashes of lightning-ignited fires that predate synthetic dioxins. In addition, a variety of simple chlorinated hydrocarbons including dichloromethane, chloroform, and carbon tetrachloride have been isolated from marine algae. A majority of the chloromethane in the environment is produced naturally by biological decomposition, forest fires, and volcanoes.

The natural organochloride epibatidine, an alkaloid isolated from tree frogs, has potent analgesic effects and has stimulated research into new pain medication. (The frogs obtain epibatidine through their diet and then sequester it on their skin. Likely dietary sources are beetles, ants, mites, and flies.)

Preparation

From Chlorine

Alkanes and aryl alkanes may be chlorinated under free radical conditions, with UV light. However, the extent of chlorination is difficult to control. Aryl chlorides may be prepared by the Friedel-Crafts halogenation, using chlorine and a Lewis acid catalyst.

The haloform reaction, using chlorine and sodium hydroxide, is also able to generate alkyl halides from methyl ketones, and related compounds. Chloroform was formerly produced thus.

Chlorine adds to the multiple bonds on alkenes and alkynes as well, giving di- or tetra-chloro compounds.

Reaction with Hydrogen Chloride

Alkenes react with hydrogen chloride (HCl) to give alkyl chlorides. For example, the industrial production of chloroethane proceeds by the reaction of ethylene with HCl:

$$H_2C = CH_2 + HCl \rightarrow CH_3CH_2Cl$$

In oxychlorination, hydrogen chloride instead of the more expensive chlorine for the same purpose:

$$CH_2 = CH_2 + 2\,HCl + \frac{1}{2}O_2 \rightarrow ClCH_2CH_2\,Cl + H_2O$$

Secondary and tertiary alcohols react with hydrogen chloride to give the corresponding chlorides. In the laboratory, the related reaction involving zinc chloride in concentrated hydrochloric acid:

$$R\text{-}OH + HCl \underset{\Delta}{\overset{ZnCl_2}{\rightarrow}} \overset{\text{alkyl}}{R} - \overset{\text{halide}}{Cl} + H_2O$$

Called the Lucas reagent, this mixture was once used in qualitative organic analysis for classifying alcohols.

Other Chlorinating Agents

Alkyl chlorides are most easily prepared by treating alcohols with thionyl chloride ($SOCl_2$) or phosphorus pentachloride (PCl_5), but also commonly with sulfuryl chloride (SO_2Cl_2) and phosphorus trichloride (PCl_3):

$$ROH + SOCl_2 \rightarrow RCl + SO_2 + HCl$$

$$3\,ROH + PCl_3 \rightarrow 3\,RCl + H_3PO_3$$

$$ROH + PCl_5 \rightarrow RCl + POCl_3 + HCl$$

In the laboratory, thionyl chloride is especially convenient, because the byproducts are gaseous. Alternatively, the Appel reaction can be used:

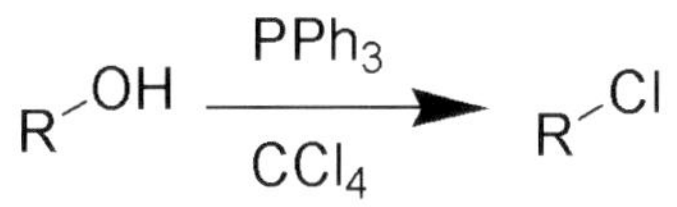

Reactions

Alkyl chlorides are versatile building blocks in organic chemistry. While alkyl bromides and iodides are more reactive, alkyl chlorides tend to be less expensive and more readily available. Alkyl chlorides readily undergo attack by nucleophiles.

Heating alkyl halides with sodium hydroxide or water gives alcohols. Reaction with alkoxides or aroxides give ethers in the Williamson ether synthesis; reaction with thiols give thioethers. Alkyl chlorides readily react with amines to give substituted amines. Alkyl chlorides are substituted by softer halides such as the iodide in the Finkelstein reaction. Reaction with other pseudohalides such as azide, cyanide, and thiocyanate are possible as well. In the presence of a strong base, alkyl chlorides undergo dehydrohalogenation to give alkenes or alkynes.

Alkyl chlorides react with magnesium to give Grignard reagents, transforming an electrophilic compound into a nucleophilic compound. The Wurtz reaction reductively couples two alkyl halides to couple with sodium.

Applications

Vinyl Chloride

The largest application of organochlorine chemistry is the production of vinyl chloride. The annual production in 1985 was around 13 billion kilograms, almost all of which was converted into polyvinylchloride (PVC).

Chloromethanes

Most low molecular weight chlorinated hydrocarbons such as chloroform, dichloromethane, dichloroethene, and trichloroethane are useful solvents. These solvents tend to be relatively non-polar; they are therefore immiscible with water and effective in cleaning applications such as degreasing and dry cleaning. Several billion kilograms of chlorinated methanes are produced annually, mainly by chlorination of methane:

$$CH_4 + XCl_2 \rightarrow CH_{4-X}\, Cl_X + XHCl$$

The most important is dichloromethane, which is mainly used as a solvent. Chloromethane is a precursor to chlorosilanes and silicones. Historically significant, but smaller in scale is chloroform, mainly a precursor to chlorodifluoromethane ($CHClF_2$) and tetrafluoroethene which is used in the manufacture of Teflon.

Pesticides

The two main groups of organochlorine insecticides are the DDT-type compounds and the chlorinated alicyclics. Their mechanism of action differs slightly.

- The DDT like compounds work on the peripheral nervous system. At the axon's sodium channel, they prevent gate closure after activation and membrane depolarization. Sodium ions leak through the nerve membrane and create a destabilizing negative "afterpotential" with hyperexcitability of the nerve. This leakage causes repeated discharges in the neuron either spontaneously or after a single stimulus.

- Chlorinated cyclodienes include aldrin, dieldrin, endrin, heptachlor, chlordane and endosulfan. A 2- to 8-hour exposure leads to depressed central nervous system (CNS) activity, followed by hyperexcitability, tremors, and then seizures. The mechanism of action is the insecticide binding at the $GABA_A$ site in the gamma-Aminobutyric acid (GABA) chloride ionophore complex, which inhibits chloride flow into the nerve.

- Other examples include dicofol, mirex, kepone, and pentachlorophenol. These can be either hydrophilic or hydrophobic, depending on their molecular structure.

Insulators

Polychlorinated biphenyls (PCBs) were once commonly used electrical insulators and heat transfer agents. Their use has generally been phased out due to health concerns. PCBs were replaced by polybrominated diphenyl ethers (PBDEs), which bring similar toxicity and bioaccumulation concerns.

Toxicity

Some types of organochlorides have significant toxicity to plants or animals, including humans. Dioxins, produced when organic matter is burned in the presence of chlorine, are persistent organic pollutantswhich pose dangers when they are released into the environment, as are some insecticides (such as DDT). For example, DDT, which was widely used to control insects in the mid-20th century, also accumulates in food chains, as do its metabolites DDE and DDD, and causes reproductive problems (e.g., eggshell thinning) in certain bird species. Some organochlorine compounds, such as sulfur mustards, nitrogen mustards, and Lewisite, are even used as chemical weapons due to their toxicity.

However, the presence of chlorine in an organic compound does not *ensure* toxicity. Some organochlorides are considered safe enough for consumption in foods and medicines. For example, peas and broad beans contain the natural chlorinated plant hormone 4-chloroindole-3-acetic acid (4-Cl-IAA); and the sweetener sucralose (Splenda) is widely used in diet products. As of 2004, at least 165 organochlorides had been approved worldwide for use as pharmaceutical drugs, including the natural antibiotic vancomycin, the antihistamine loratadine (Claritin), the antidepressant sertraline (Zoloft), the anti-epileptic lamotrigine (Lamictal), and the inhalation anesthetic isoflurane.

Rachel Carson brought the issue of DDT pesticide toxicity to public awareness with her 1962 book *Silent Spring*. While many countries have phased out the use of some types of organochlorides such as the US ban on DDT, persistent DDT, PCBs, and other organochloride residues continue to be found in humans and mammals across the planet many years after production and use have been limited. In Arcticareas, particularly high levels are found in marine mammals. These chemicals concentrate in mammals, and are even found in human breast milk. In some species of marine mammals, particularly those that produce milk with a high fat content, males typically have far higher levels, as females reduce their concentration by transfer to their offspring through lactation.

Exposure to Organochlorine Pesticides

People can be exposed to organochlorine pesticides through accidental inhalation exposure if you are in an area where they were recently applied. The chemicals can also be ingested in fish, dairy products, and other fatty foods that are contaminated. Organochlorine pesticides accumulate in the environment. They are very persistent and move long distances in surface runoff or groundwater. Prior to the mid-1970s, organochlorines resulted in widespread reproductive failure among birds because birds laid eggs with thin shells that cracked before hatching.

Effects of Pesticides on Humans

Exposure to organochlorine pesticides over a short period may produce convulsions, headache, dizziness, nausea, vomiting, tremors, confusion, muscle weakness, slurred speech, salivation and sweating. Long-term exposure to organochlorine pesticides may damage the liver, kidney, central nervous system, thyroid and bladder. Many of these pesticides have been linked to elevated rates of liver or kidney cancer in animals. There is some evidence indicating that organochlorine pesticides may also cause cancer in humans.

DDT

DDT, abbreviation of dichlorodiphenyltrichloroethane, also called 1,1,1-trichloro-2,2-bis (*p*-chlorophenyl)ethane, a synthetic insecticide belonging to the family of organic halogen compounds, highly toxic toward a wide variety of insects as a contact poison that apparently exerts its effect by disorganizing the nervous system.

DDT, prepared by the reaction of chloral with chlorobenzene in the presence of sulfuric acid, was first made in 1874; its insecticidal properties were discovered in 1939 by a Swiss chemist, Paul Hermann Müller. During and after World War II, DDT was found to be effective against lice, fleas, and mosquitoes (the carriers of typhus, of plague, and of malaria and yellow fever, respectively) as well as the Colorado potato beetle, the gypsy moth, and other insects that attack valuable crops.

Many species of insects rapidly develop populations resistant to DDT; the high stability of the compound leads to its accumulation in insects that constitute the diet of other animals, with toxic effects on them, especially certain birds and fishes. These two disadvantages had severely decreased the value of DDT as an insecticide by the 1960s, and severe restrictions were imposed on its use in the United States in 1972.

Pure DDT is a colourless, crystalline solid that melts at 109° C (228° F); the commercial product, which is usually 65 to 80 percent active compound, along with related substances, is an amorphous powder that has a lower melting point. DDT is applied as a dust or by spraying its aqueous suspension.

Properties and Chemistry

DDT is similar in structure to the insecticide methoxychlor and the acaricide dicofol. It is highly hydrophobic and nearly insoluble in water but has good solubility in most organic solvents, fats and oils. DDT does not occur naturally and is synthesised by a Friedel–Crafts hydroxyalkylation reaction between chloral (CCl_3CHO) and chlorobenzene (C_6H_5Cl), in the presence of an acidic catalyst. DDT has been marketed under trade names including Anofex, Cezarex, Chlorophenothane, Clofenotane, Dicophane, Dinocide, Gesarol, Guesapon, Guesarol, Gyron, Ixodex, Neocid, Neocidol and Zerdane.

Isomers and Related Compounds

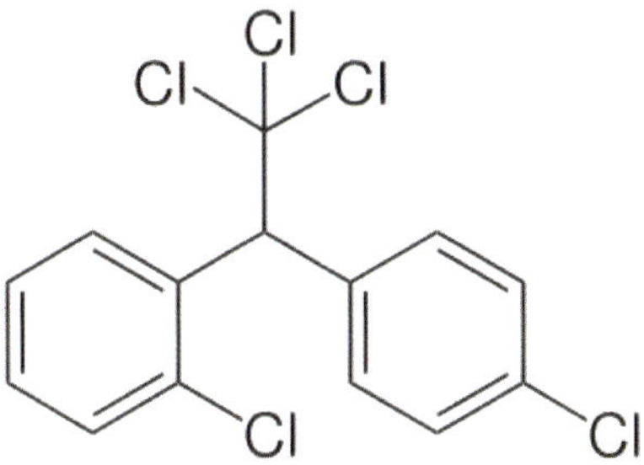

o,p'-DDT, a minor component in commercial DDT.

Commercial DDT is a mixture of several closely–related compounds. The major component (77%) is the *p,p'* isomer (pictured above). The *o,p'* isomer (pictured to the right) is also present in significant amounts (15%). Dichlorodiphenyldichloroethylene (DDE) and dichlorodiphenyldichloroethane (DDD) make up the balance. DDE and DDD are the major metabolites and environmental breakdown products.

Production and use

DDT has been formulated in multiple forms, including solutions in xylene or petroleum distillates, emulsifiable concentrates, water-wettable powders, granules, aerosols, smoke candles and charges for vaporizers and lotions.

From 1950 to 1980, DDT was extensively used in agriculture – more than 40,000 tonnes each year worldwide – and it has been estimated that a total of 1.8 million tonnes have been produced globally since the 1940s. In the United States, it was manufactured by some 15 companies, including Monsanto, Ciba,Montrose Chemical Company, Pennwalt, and Velsicol Chemical Corporation. Production peaked in 1963 at 82,000 tonnes per year. More than 600,000 tonnes (1.35 billion pounds) were applied in the US before the 1972 ban. Usage peaked in 1959 at about 36,000 tonnes.

In 2009, 3,314 tonnes were produced for malaria control and visceral leishmaniasis. India is the only country still manufacturing DDT, and is the largest consumer. China ceased production in 2007.

Mechanism of Insecticide Action

In insects, DDT opens sodium ion channels in neurons, causing them to fire spontaneously, which leads to spasms and eventual death. Insects with certain mutations in their sodium channel gene are resistant to DDT and similar insecticides. DDT resistance is also conferred by up-regulation of genes expressing cytochrome P450 in some insect species, as greater quantities of some enzymes of this group accelerate the toxin›s metabolism into inactive metabolites. (The same enzyme family is up-regulated in mammals too, e.g., in response to ethanol consumption.) Genomic studies in the model genetic organism *Drosophila melanogaster* revealed that high level DDT resistance is polygenic, involving multiple resistance mechanisms.

Environmental Impact

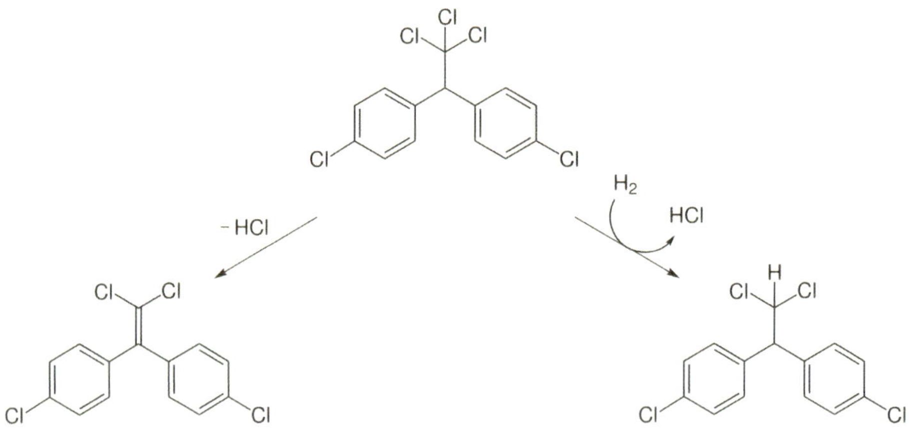

Degradation of DDT to form DDE (by elimination of HCl, left) and DDD (by reductive dechlorination, right)

DDT is a persistent organic pollutant that is readily adsorbed to soils and sediments, which can act both as sinks and as long-term sources of exposure affecting organisms. Depending on conditions, its soil half-life can range from 22 days to 30 years. Routes of loss and degradation include runoff, volatilization, photolysis and aerobic and anaerobic biodegradation. Due to hydrophobic properties, in aquatic ecosystems DDT and its metabolites are absorbed by aquatic organisms and adsorbed on suspended particles, leaving little DDT dissolved in the water (however, its half-life in aquatic environments is listed by the National Pesticide Information Center as 150 years). Its breakdown products and metabolites, DDE and DDD, are also persistent and have similar chemical and physical properties. DDT and its breakdown products are transported from warmer areas to the Arctic by the phenomenon of global distillation, where they then accumulate in the region's food web.

Medical researchers in 1974 found a measurable and significant difference in the presence of DDT in human milk between mothers who lived in New Brunswick and mothers who lived in Nova Scotia, "possibly because of the wider use of insecticide sprays in the past."

Because of its lipophilic properties, DDT can bioaccumulate, especially in predatory birds. DDT is toxic to a wide range of living organisms, including marine animals such as crayfish, daphnids, sea shrimp and many species of fish. DDT, DDE and DDD magnify through the food chain, with apex predators such as raptor birds concentrating more chemicals than other animals in the same environment. They are stored mainly in body fat. DDT and DDE are resistant to metabolism; in humans, their half-lives are 6 and up to 10 years, respectively. In the United States, these chemicals were detected in almost all human blood samples tested by the Centers for Disease Controlin 2005, though their levels have sharply declined since most uses were banned. Estimated dietary intake has declined, although FDA food tests commonly detect it.

Eggshell Thinning

The chemical and its breakdown products DDE and DDD caused eggshell thinning and population declines in multiple North American and European bird of prey species. DDE-related eggshell thinning is considered a major reason for the decline of the bald eagle, brown pelican, peregrine falcon and osprey. However, birds vary in their sensitivity to these chemicals, with birds of prey, waterfowl and song birds being more susceptible than chickens and related species. Even in 2010, California condors that feed on sea lions at Big Sur that in turn feed in the Palos Verdes Shelf area of the Montrose Chemical Superfund site exhibited continued thin-shell problems, though DDT›s role in the decline of the California condor is disputed.

The biological thinning mechanism is not entirely understood, but DDE appears to be more potent than DDT, and strong evidence indicates that p,p›-DDE inhibits calcium ATPase in the membrane of the shell gland and reduces the transport of calcium carbonate from blood into the eggshell gland. This results in a dose-dependent thickness reduction. Other evidence indicates that o,p›-DDT disrupts female reproductive tract development, later impairing eggshell quality. Multiple mechanisms may be at work, or different mechanisms may operate in different species.

Human Health

DDT is an endocrine disruptor. It is considered likely to be a human carcinogen although the majority of studies suggest it is not directly genotoxic. DDEacts as a weak androgen receptor antagonist, but not as an estrogen. p,p'-DDT, DDT's main component, has little or no androgenic or estrogenic activity. The minor component o,p'-DDT has weak estrogenic activity.

A U.S. soldier is demonstrating DDT hand-spraying equipment. DDT was used to control the spread of typhus-carrying lice.

Acute Toxicity

DDT is classified as "moderately toxic" by the US National Toxicology Program (NTP) and "moderately hazardous" by WHO, based on the rat oral LD_{50} of 113 mg/kg. Indirect exposure is considered relatively non-toxic for humans.

Chronic Toxicity

Primarily through the tendency for DDT to buildup in areas of the body with high lipid content, chronic exposure can affect reproductive capabilities and the embryo or fetus.

- Research has shown that exposure to DDT at amounts that would be needed in malaria control might cause preterm birth and early weaning; toxicological evidence shows endocrine-disrupting properties; human data also indicate possible disruption in semen quality, menstruation, gestational length, and duration of lactation.
- Other studies document decreases in semen quality among men with high exposures (generally from IRS).
- Studies are inconsistent on whether high blood DDT or DDE levels increase time to pregnancy. In mothers with high DDE blood serum levels, daughters may have up to a 32% increase in the probability of conceiving, but increased DDT levels have been associated with a 16% decrease in one study.
- Indirect exposure of mothers through workers directly in contact with DDT is associated with an increase in spontaneous abortions.

- Other studies found that DDT or DDE interfere with proper thyroid function in pregnancy and childhood.

Carcinogenicity

In 2015, the International Agency for Research on Cancer classified DDT as Group 2A "probably carcinogenic to humans". Previous assessments by the U.S. National Toxicology Program classified it as "reasonably anticipated to be a carcinogen" and by the EPA classified DDT, DDE and DDD as class B2 "probable" carcinogens; these evaluations were based mainly on animal studies.

Occupational DDT exposure was associated with increased pancreatic cancer risk in 2 case control studies, but another study showed no DDE dose-effect association. Results regarding a possible association with liver cancer and biliary tract cancer are conflicting: workers who did not have direct occupational DDT contact showed increased risk. White men had an increased risk, but not white women or black men. Results about an association with multiple myeloma, prostate and testicular cancer, endometrial cancer and colorectal cancer have been inconclusive or generally do not support an association. A 2017 review of liver cancer studies concluded that "organochlorine pesticides, including DDT, may increase hepatocellular carcinoma risk."

A 2009 review, whose co-authors included persons engaged in DDT-related litigation, reached broadly similar conclusions, with an equivocal association with testicular cancer. Case–control studies did not support an association with leukemia or lymphoma.

Breast Cancer

The question of whether DDT or DDE are risk factors in breast cancer has not been conclusively answered. Several meta analyses of observational studies have concluded that there is no overall relationship between DDT exposure and breast cancer risk. The United States Institute of Medicine reviewed data on the association of breast cancer with DDT exposure in 2012 and concluded that a causative relationship could neither be proven nor disproven.

A 2007 case control study using archived blood samples found that breast cancer risk was increased 5-fold among women who were born prior to 1931 and who had high serum DDT levels in 1963. Reasoning that DDT use became widespread in 1945 and peaked around 1950, they concluded that the ages of 14–20 were a critical period in which DDT exposure leads to increased risk. This study, which suggests a connection between DDT exposure and breast cancer that would not be picked up by most studies, has received variable commentary in third party reviews. One review suggested that "previous studies that measured exposure in older women may have missed the critical period." The National Toxicology Program notes that while the majority of studies have not found a relationship between DDT exposure and breast cancer that positive associations have been seen in a "few studies among women with higher levels of exposure and among certain subgroups of women"

A 2015 case control study identified a link (odds ratio 3.4) between *in-utero* exposure (as estimated from archived maternal blood samples) and breast cancer diagnosis in daughters. The findings "support classification of DDT as an endocrine disruptor, a predictor of breast cancer, and a marker of high risk".

Malaria Control

Malaria remains the primary public health challenge in many countries. In 2015, there were 214 million cases of malaria worldwide resulting in an estimated 438,000 deaths, 90% of which occurred in Africa.DDT is one of many tools to fight the disease. Its use in this context has been called everything from a "miracle weapon [that is] like Kryptonite to the mosquitoes," to "toxic colonialism".

Before DDT, eliminating mosquito breeding grounds by drainage or poisoning with Paris green or pyrethrum was sometimes successful. In parts of the world with rising living standards, the elimination of malaria was often a collateral benefit of the introduction of window screens and improved sanitation. A variety of usually simultaneous interventions represents best practice. These include antimalarial drugs to prevent or treat infection; improvements in public health infrastructure to diagnose, sequester and treat infected individuals; bednets and other methods intended to keep mosquitoes from biting humans; and vector control strategies such as larvaciding with insecticides, ecological controls such as draining mosquito breeding grounds or introducing fish to eat larvae and indoor residual spraying (IRS) with insecticides, possibly including DDT. IRS involves the treatment of interior walls and ceilings with insecticides. It is particularly effective against mosquitoes, since many species rest on an indoor wall before or after feeding. DDT is one of 12 WHO–approved IRS insecticides.

WHO's anti-malaria campaign of the 1950s and 1960s relied heavily on DDT and the results were promising, though temporary in developing countries. Experts tie malarial resurgence to multiple factors, including poor leadership, management and funding of malaria control programs; poverty; civil unrest; and increased irrigation. The evolution of resistance to first-generation drugs (e.g. chloroquine) and to insecticides exacerbated the situation. Resistance was largely fueled by unrestricted agricultural use. Resistance and the harm both to humans and the environment led many governments to curtail DDT use in vector control and agriculture. In 2006 WHO reversed a long-standing policy against DDT by recommending that it be used as an indoor pesticide in regions where malaria is a major problem.

Once the mainstay of anti-malaria campaigns, as of 2008 only 12 countries used DDT, including India and some southern African states, though the number was expected to rise.

Initial Effectiveness

When it was introduced in World War II, DDT was effective in reducing malaria morbidity and mortality. WHO's anti-malaria campaign, which consisted mostly of spraying DDT and rapid treatment and diagnosis to break the transmission cycle, was initially successful as well. For example, in Sri Lanka, the program reduced cases from about one million per year before spraying to just 18 in 1963 and 29 in 1964. Thereafter the program was halted to save money and malaria rebounded to 600,000 cases in 1968 and the first quarter of 1969. The country resumed DDT vector control but the mosquitoes had evolved resistance in the interim, presumably because of continued agricultural use. The program switched to malathion, but despite initial successes, malaria continued its resurgence into the 1980s.

DDT remains on WHO's list of insecticides recommended for IRS. After the appointment of Arata Kochi as head of its anti-malaria division, WHO's policy shifted from recommending IRS only in areas of seasonal or episodic transmission of malaria, to advocating it in areas of continuous, intense transmission. WHO reaffirmed its commitment to phasing out DDT, aiming "to achieve a 30% cut in the application of DDT world-wide by 2014 and its total phase-out by the early 2020s if not sooner" while simultaneously combating malaria. WHO plans to implement alternatives to DDT to achieve this goal.

South Africa continues to use DDT under WHO guidelines. In 1996, the country switched to alternative insecticides and malaria incidence increased dramatically. Returning to DDT and introducing new drugs brought malaria back under control. Malaria cases increased in South America after countries in that continent stopped using DDT. Research data showed a strong negative relationship between DDT residual house sprayings and malaria. In a research from 1993 to 1995, Ecuador increased its use of DDT and achieved a 61% reduction in malaria rates, while each of the other countries that gradually decreased its DDT use had large increases.

Mosquito Resistance

In some areas resistance reduced DDT's effectiveness. WHO guidelines require that absence of resistance must be confirmed before using the chemical. Resistance is largely due to agricultural use, in much greater quantities than required for disease prevention.

Resistance was noted early in spray campaigns. Paul Russell, former head of the Allied Anti-Malaria campaign, observed in 1956 that "resistance has appeared after six or seven years." Resistance has been detected in Sri Lanka, Pakistan, Turkey and Central America and it has largely been replaced by organophosphate or carbamate insecticides, *e.g.* malathion or bendiocarb.

In many parts of India, DDT is ineffective. Agricultural uses were banned in 1989 and its anti-malarial use has been declining. Urban use ended. One study concluded that

"DDT is still a viable insecticide in indoor residual spraying owing to its effectivity in well supervised spray operation and high excito-repellency factor."

Studies of malaria-vector mosquitoes in KwaZulu-Natal Province, South Africa found susceptibility to 4% DDT (WHO›s susceptibility standard), in 63% of the samples, compared to the average of 87% in the same species caught in the open. The authors concluded that "Finding DDT resistance in the vector *An. arabiensis*, close to the area where we previously reported pyrethroid-resistance in the vector *An. funestus* Giles, indicates an urgent need to develop a strategy of insecticide resistance management for the malaria control programmes of southern Africa."

DDT can still be effective against resistant mosquitoes and the avoidance of DDT-sprayed walls by mosquitoes is an additional benefit of the chemical. For example, a 2007 study reported that resistant mosquitoes avoided treated huts. The researchers argued that DDT was the best pesticide for use in IRS (even though it did not afford the most protection from mosquitoes out of the three test chemicals) because the others pesticides worked primarily by killing or irritating mosquitoes – encouraging the development of resistance. Others argue that the avoidance behavior slows eradication. Unlike other insecticides such as pyrethroids, DDT requires long exposure to accumulate a lethal dose; however its irritant property shortens contact periods. "For these reasons, when comparisons have been made, better malaria control has generally been achieved with pyrethroids than with DDT." In India outdoor sleeping and night duties are common, implying that "the excito-repellent effect of DDT, often reported useful in other countries, actually promotes outdoor transmission."

Residents' Concerns

IRS is effective if at least 80% of homes and barns in a residential area are sprayed. Lower coverage rates can jeopardize program effectiveness. Many residents resist DDT spraying, objecting to the lingering smell, stains on walls, and the potential exacerbation of problems with other insect pests. Pyrethroid insecticides (e.g. deltamethrin and lambda-cyhalothrin) can overcome some of these issues, increasing participation.

Human Exposure

A 1994 study found that South Africans living in sprayed homes have levels that are several orders of magnitude greater than others. Breast milk from South African mothers contains high levels of DDT and DDE. It is unclear to what extent these levels arise from home spraying vs food residues. Evidence indicates that these levels are associated with infant neurological abnormalities.

Most studies of DDT's human health effects have been conducted in developed countries where DDT is not used and exposure is relatively low.

Illegal diversion to agriculture is also a concern as it is difficult to prevent and its subsequent use on crops is uncontrolled. For example, DDT use is widespread in Indian agriculture, particularly mangoproduction and is reportedly used by librarians to protect books. Other examples include Ethiopia, where DDT intended for malaria control is reportedly used in coffee production, and Ghana where it is used for fishing." The residues in crops at levels unacceptable for export have been an important factor in bans in several tropical countries. Adding to this problem is a lack of skilled personnel and management.

Criticism of Restrictions on DDT use

A few people and groups have argued that limitations on DDT use for public health purposes have caused unnecessary morbidity and mortality from vector-borne diseases, with some claims of malaria deaths ranging as high as the hundreds of thousands and millions. Robert Gwadz of the US National Institutes of Health said in 2007, "The ban on DDT may have killed 20 million children." These arguments were rejected as "outrageous" by former WHO scientist Socrates Litsios. May Berenbaum, University of Illinois entomologist, says, "to blame environmentalists who oppose DDT for more deaths than Hitler is worse than irresponsible."

Criticisms of a DDT "ban" often specifically reference the 1972 United States ban (with the erroneous implication that this constituted a worldwide ban and prohibited use of DDT in vector control). Reference is often made to *Silent Spring,* even though Carson never pushed for a DDT ban. John Quiggin and Tim Lambert wrote, "the most striking feature of the claim against Carson is the ease with which it can be refuted."

Alternatives

Insecticides

Organophosphate and carbamate insecticides, *e.g.* malathion and bendiocarb, respectively, are more expensive than DDT per kilogram and are applied at roughly the same dosage. Pyrethroids such as deltamethrin are also more expensive than DDT, but are applied more sparingly (0.02–0.3 g/m^2 vs 1–2 g/m^2), so the net cost per house is about the same.

Non-chemical Vector Control

Before DDT, malaria was successfully eliminated or curtailed in several tropical areas by removing or poisoning mosquito breeding grounds and larva habitats, for example by eliminating standing water. These methods have seen little application in Africa for more than half a century. According to CDC, such methods are not practical in Africa because "*Anopheles gambiae*, one of the primary vectors of malaria in Africa, breeds in numerous small pools of water that form due to rainfall. It is difficult, if not impossible, to predict when and where the breeding sites will form, and to find and treat them before the adults emerge."

The relative effectiveness of IRS versus other malaria control techniques (e.g. bednets or prompt access to anti-malarial drugs) varies and is dependent on local conditions.

A WHO study released in January 2008 found that mass distribution of insecticide-treated mosquito nets and artemisinin–based drugs cut malaria deaths in half in malaria-burdened Rwanda and Ethiopia. IRS with DDT did not play an important role in mortality reduction in these countries.

Vietnam has enjoyed declining malaria cases and a 97% mortality reduction after switching in 1991 from a poorly funded DDT-based campaign to a program based on prompt treatment, bednets and pyrethroid group insecticides.

In Mexico, effective and affordable chemical and non-chemical strategies were so successful that the Mexican DDT manufacturing plant ceased production due to lack of demand.

A review of fourteen studies in sub-Saharan Africa, covering insecticide-treated nets, residual spraying, chemoprophylaxis for children, chemoprophylaxis or intermittent treatment for pregnant women, a hypothetical vaccine and changing front–line drug treatment, found decision making limited by the lack of information on the costs and effects of many interventions, the small number of cost-effectiveness analyses, the lack of evidence on the costs and effects of packages of measures and the problems in generalizing or comparing studies that relate to specific settings and use different methodologies and outcome measures. The two cost-effectiveness estimates of DDT residual spraying examined were not found to provide an accurate estimate of the cost-effectiveness of DDT spraying; the resulting estimates may not be good predictors of cost-effectiveness in current programs.

However, a study in Thailand found the cost per malaria case prevented of DDT spraying (US$1.87) to be 21% greater than the cost per case prevented of lambda-cyhalothrin–treated nets (US$1.54),casting some doubt on the assumption that DDT was the most cost-effective measure. The director of Mexico's malaria control program found similar results, declaring that it was 25% cheaper for Mexico to spray a house with synthetic pyrethroids than with DDT. However, another study in South Africa found generally lower costs for DDT spraying than for impregnated nets.

A more comprehensive approach to measuring cost-effectiveness or efficacy of malarial control would not only measure the cost in dollars, as well as the number of people saved, but would also consider ecological damage and negative human health impacts. One preliminary study found that it is likely that the detriment to human health approaches or exceeds the beneficial reductions in malarial cases, except perhaps in epidemics. It is similar to the earlier study regarding estimated theoretical infant mortality caused by DDT.

A study in the Solomon Islands found that "although impregnated bed nets cannot entirely replace DDT spraying without substantial increase in incidence, their use permits reduced DDT spraying."

A comparison of four successful programs against malaria in Brazil, India, Eritrea and Vietnam does not endorse any single strategy but instead states, "Common success factors included conducive country conditions, a targeted technical approach using a package of effective tools, data-driven decision-making, active leadership at all levels of government, involvement of communities, decentralized implementation and control of finances, skilled technical and managerial capacity at national and sub-national levels, hands-on technical and programmatic support from partner agencies, and sufficient and flexible financing."

DDT resistant mosquitoes have generally proved susceptible to pyrethroids. Thus far, pyrethroid resistance in *Anopheles* has not been a major problem.

DDT Accumulation

Fat Stores

- DDT tends to accumulate in the fatty tissues of insects, wildlife, and people, but produces no known toxic effects while it is stored in the fat.
- DDT is metabolized into various breakdown products in the body including DDE, DDD[4] , and DDA[5].
- When fat stores are used during periods of starvation the breakdown products of DDT are released into the blood where they may be toxic to the liver and the nervous system.
- Once DDT has accumulated in the body, it is excreted in the urine, feces, or breast milk. Breast milk is often used to measure a population's exposure to DDT.

Biomagnification

- Because of DDT's chemical properties it has the tendency to accumulate in animals. As animals lower on the food chain are eaten by other animals higher up, DDT becomes concentrated in the fatty tissues of predators. This continues until the primary predator of the food chain receives the highest dose, which may lead to adverse health effects. Once the use of DDT was discontinued in the U.S., its concentration in the environment and animals decreased.

Aldrin

Aldrin is a pesticide used to control soil insects such as termites, corn rootworm, wireworms, rice water weevil, and grasshoppers. It has been widely used to protect crops such as corn and potatoes, and has been effective to protect wooden structures from termites.

Aldrin is readily metabolized to dieldrin by both plants and animals. As a result, Aldrin residues are rarely found in foods and animals, and then only in small amounts. It binds strongly to soil particles and is very resistant to leaching into groundwater. Volatilization is an important mechanism of loss from soils. Due to its persistent nature and hydrophobicity, Aldrin is known to bioconcentrate, mainly as its conversion products.

Aldrin is banned in many countries, including Bulgaria, Ecuador, Finland, Hungary, Israel, Singapore, Switzerland, and Turkey. Its use is severely restricted in many countries, including Argentina, Austria, Canada, Chile, the EU, Japan, New Zealand, the Philippines, USA, and Venezuela.

Use in South East Asia

	Used or Found in Country?	Years of Usage	Regulatory Controls
Cambodia			Banned in 1992
Lao PDR			Banned in 1992
Malaysia	Yes	1980-1994	Banned in 1994
Thailand	Yes	1955	Banned in 1988
Indonesia			Never registered for use
Philippines			Banned in 1989
Viet Nam			Banned in 1992

Effects on Plants and Animals

Aldrin has low phytotoxicity with plants affected only by extremely high application rates. The toxicity of Aldrin to aquatic organisms is quite variable, with aquatic insects being the most sensitive group of invertebrates. The 96-h LC50 dose values range from 1-200 μg/L for insects, and from 2.2-53 μg/L for fish. Long term and bioconcentration studies are performed primarily using Dieldrin, the primary conversion product of Aldrin. In a model ecosystem study, only 0.5% of the original radioactive Aldrin was stored as Aldrin in the mosquitofish (Gambusia affinis), the organism at the top of the model food chain.

The acute toxicity of Aldrin to avian species varies in the range of 6.6 mg/kg for bobwhite quail to 520 mg/kg for mallard ducks. Aldrin treated rice is thought to have been the cause of deaths in waterfowl, shorebirds and passerines along the Texas Gulf Coast, both by direct poisoning by ingestion of Aldrin treated rice and indirectly by consuming organisms contaminated with Aldrin. Residues of Aldrin were detected in all samples of bird casualties, eggs, scavengers, predators, fish, frogs, invertebrates and soil.

As Aldrin is readily and rapidly converted to Dieldrin in the environment its fate is closely linked to that of Dieldrin. Aldrin is readily metabolised to dieldrin in both animals and plants, and therefore Aldrin residues are rarely present in animals and then only in very small amounts. Residues of Aldrin have been detected in fish in Egypt, the average concentration was 8.8 µg/kg, and a maximum concentration of 54.27 µg/kg.

Effects on Humans

Aldrin is toxic to humans; the lethal dose of Aldrin for an adult man has been estimated to be about 5g, equivalent to 83 mg/kg body weight. Signs and symptoms of Aldrin intoxication may include headache, dizziness, nausea, general malaise and vomiting, followed by muscle twitching, myoclonic jerks, and convulsions. Occupational exposure to Aldrin, in conjunction with Dieldrin and Endrin, was associated with a significant increase in liver and biliary cancer, although the study did have some limitations, including a lack of quantitative exposure information. There is limited information that cyclodienes, such as Aldrin, may affect immune responses.

The acute oral LD50 dose for Aldrin in laboratory animals is in the range of 33 mg/kg body weight for guinea pigs to 320 mg/kg body weight for hamsters. Reproductive effects in rats were observed when pregnant females were dosed with 1.0 mg/kg Aldrin subcutaneously. Offspring experienced a decrease in the median effective time for incisor teeth eruption and an increase in the median effective time for testes descent. There is, as yet, no evidence of a teratogenic potential for Aldrin. The International Agency for Research on Cancer (IARC) has concluded that there is inadequate evidence for the carcinogenicity of Aldrin in humans, and there is only limited evidence in experimental animals. Aldrin is therefore not classifiable as to its carcinogenicity in humans.

The average daily human intake of Aldrin and Dieldrin was calculated to be 19µg/person in India, and 0.55 µg/person in Vietnam. Dairy products, such as milk and butter, and animal meats are the primary sources of exposure.

Endrin

Endrin is a foliar insecticide used mainly on field crops such as cotton and grains. It has also been used as a rodenticide to control mice and voles. It is rapidly metabolised by animals and does not accumulate in fat to the same extent as other compounds with similar structures. It can enter the atmosphere by volatilization, and can contaminate surface water from soil run-off.

Endrin is banned in many countries, including Belgium, Cyprus, Ecuador, Finland, Israel, Philippines, Singapore, Thailand, and Togo. Its use is severely restricted in many countries, including Argentina, Canada, Chile, Colombia, the EU, India, Japan, New Zealand, Pakistan, USA, and Venezuela.

Usage in South East Asia

	Used or Found in Country?	Years of Usage	Regulatory Controls
Cambodia	Yes	1950-92 Present?	Banned
Lao PDR			Banned in 1992
Malaysia			Withdrawn by company
Thailand	Yes	1955	Banned in 1981
Indonesia			Never registered for use
Philippines			Banned from import in 1989
Viet Nam			Banned in 1992

Potential Effects on Humans

A study of workers involved in the production of Aldrin, Dieldrin and Endrin did not find Endrin in the blood of workers, except in cases of accidental, acute over-exposure. These findings are in agreement with results of a study of 71 workers in an Endrin plant in the USA. Data on absenteeism, results of liver function tests, blood chemistry, blood morphology, urine analysis, occurrence of sensitization, and the incidence and pattern of diseases, including the occurrence of malignant growth, showed no difference between workers exposed to Endrin and other chemical plant operators. A study of workers involved in the manufacture of Aldrin, Dieldrin and Endrin found a statistically significant increase in liver and biliary tract cancers, although the study did have some limitations such as a lack of quantitative exposure information. There is limited evidence that cyclodienes such as Endrin may also depress immune responses.

IARC has concluded that there is inadequate evidence for the carcinogenicity of Endrin in humans, and there is only limited evidence in experimental animals. Endrin is therefore not classifiable as to its carcinogenicity in humans.

Potential Effects on Animals

Endrin is metabolised rapidly by animals, and very little is accumulated in fat compared to compounds of similar structure (including its stereoisomer - Dieldrin). The formation of anti-12-hydroxyEndrin is considered to be the major route of metabolism of Endrin.

The acute oral LD50 of Endrin is in the range of 3 mg/kg body weight in monkeys to 36 mg/kg in guinea pigs.

Male and female Long-Evans rats were fed Endrin in their diet over three generations. No difference in appearance, behaviour, body weight, or number or size of litters was observed. The weights of liver, kidneys and brain were normal, and no histopathological abnormalities were observed in third generation weanlings.

Significant increased mortality of pups in the second and third generations of rats fed 3 mg/kg was noted. Endrin was not teratogenic at levels that did not cause maternal toxicity.

Endrin is highly toxic to fish, with most LC50 values below 1.0 μg/L. Sheepshead minnow embryos exposed for 23 weeks to 0.31 and 0.72 μg/L hatched early, and all those exposed to 0.72 μg/L died by the ninth day of their exposure, while those exposed at 0.31 μg/L were initially stunted and some died. The reproductive ability of the survivors of the 0.31 μg/L was impaired. No significant effects were observed at an exposure concentration of 0.12 μg/L. The lowest observed adverse effect level (LOAEL) for aquatic organisms was 30 ng/L over 20 days for reproduction in mysid shrimp. Reproduction in male and female mallard ducks was not impaired by diets containing 0, 0.5 or 3.0 mg/kg.

Environmental Behavior

Insecticides like dieldrin and endrin have been shown to persist for decades in the environment. A definitive detection of the residues was not possible until 1971 when mass spectrometer started being used as a detector in gas chromatography. Detection of these chemicals in the environment has been reported across the world up to 2005, even though the frequency of reported cases are low due to its relatively small-scale use and very low concentrations.

Endrin regularly enters the environment when applied to crops or when rain washes it off. It has been found in water, sediments, atmospheric air and biotic environment, even after uses have been stopped.Organochlorine pesticides strongly resist degradation, are poorly soluble in water but highly soluble in lipids, which is called lipophilic. This leads to bioaccumulation in fatty tissues of organisms, mainly those dwelling in water. A high bioconcentration factor of 1335-10,000 has been reported in fish. Endrin binds very strongly to organic matter in soil and aquatic sediments due to their high adsorption coefficient,making it less likely to leach into groundwater, even though contaminated groundwater samples have been found. In 2009, EPA released data indicating that the endrin in soil could last up to 14 years or more. The extent of endrin›s persistence depends highly on local conditions. For example, high temperature (230 °C) or intense sunlight leads to more rapid breakdown of endrin into endrin ketone and endrin aldehyde, however, this breakdown is less than 5%.

Removal from the Environment

In the United States, endrin was mainly disposed in land until U.S. federal regulations were applied in 1987 on land disposal of wastes containing endrin. Primary methods of endrin disappearance from soil are volatilization and photodecomposition. Under ultraviolet light, endrin forms δ-ketoendrin and International Programme on Chemical

Safety (IPCS) claims that in intense summer sun, about 50% of endrin is isomerized to δ-ketoendrin in 7 days. In anaerobic conditions microbial degradation by fungi and bacteria takes place to form the same major end product.

Hazardous Substances Data Bank (HSDB) lists reductive dechlorination and incineration for field disposal of small quantities of endrin. In reductive dechlorination, endrin›s chlorine atoms were completely replaced with hydrogen atoms, which is suspected to be more environmentally acceptable. Even though endrin binds very strongly to soil, phytoremediation has been proposed by group of Japanese scientists using crops in the Cucurbitaceae family. As of 2009, exact mechanisms behind the plant uptake of endrin have not been understood. Research in uptake mechanisms and factors that influence the uptake is needed for practical application.

Heptachlor

Heptachlor was used as an insecticide; however, nearly all registered uses of heptachlor have been canceled. Very limited information is available on the health effects of heptachlor in humans. Acute (short-term) inhalation exposure to heptachlor may result in nervous system effects, with oral studies showing gastrointestinal effects. Animal studies have reported effects on the liver and central nervous system from oral exposure. Chronic (long-term) inhalation and oral exposure by humans may be associated with neurological effects including irritability, salivation, and dizziness, while oral exposure may result in effects on the blood. Human studies are inconclusive regarding heptachlor and cancer. Animal oral studies have reported liver tumors. EPA has classified heptachlor as a Group B2, probable human carcinogen.

Uses

- Heptachlor is a constituent of technical grade chlordane, approximately 10 percent by weight.
- Heptachlor was used as an insecticide in the United States from 1953 to 1974. In 1974, nearly all registered uses of heptachlor were canceled.
- Heptachlor was used from 1953 to 1974 as a soil and seed treatment to protect corn, small grains, and sorghum from pests. It was also used to control ants, cutworms, maggots, termites, and other pests in agriculture and in the home.
- The sale of heptachlor was voluntarily canceled in 1987 by its sole U.S. manufacturer. In 1988, the sale, distribution, and shipment of existing stocks of all canceled heptachlor and chlordane products was prohibited in the United States.
- The only commercial use of heptachlor products still permitted is fire ant control in power transformers. In addition, homeowner's use of existing stocks of heptachlor-containing termite control products is also allowed.

Sources and Potential Exposure

- People whose homes were treated for termites with heptachlor may be exposed to heptachlor in the indoor air for many years after treatment. Workers who use heptachlor to kill fire ants or who manufacture the chemical may be exposed to it in the air or through the skin.
- Heptachlor has been detected in food, including fish, shellfish, dairy products, meat, and poultry. Another possible source of exposure is drinking water; heptachlor has been detected at low concentrations in drinking water wells in several states.

Assessing Personal Exposure

- Laboratory tests can detect heptachlor (and its breakdown product, heptachlor epoxide) in blood, fat, breast milk, and body tissues after exposure to high levels.

Health Hazard Information

Acute Effects:

- Acute inhalation exposure to heptachlor in humans has been associated with nervous system effects in a few case studies, while gastrointestinal effects, such as nausea and vomiting, have been reported to occur following accidental ingestion of heptachlor.
- Effects on the liver and central nervous system have been noted in animals acutely exposed to heptachlor via the oral route.
- Heptachlor is considered to have high to extreme acute toxicity based on short-term oral tests in rats.

Chronic Effects (Noncancer):

- Chronic inhalation exposure to heptachlor has been associated with blood effects in humans, while oral exposure has resulted in neurological effects including irritability, salivation, dizziness, muscle tremors, and convulsions.
- Animal studies have reported effects on the liver, kidney, and the immune and nervous systems from oral exposure to heptachlor.
- The Reference Dose (RfD) for heptachlor is 0.0005 milligrams per kilogram body weight per day (mg/kg/d) based on liver weight increases in males rats only. The RfD is an estimate (with uncertainty spanning perhaps an order of magnitude) of a daily oral exposure to the human population (including sensitive subgroups) that is likely to be without appreciable risk of deleterious

noncancer effects during a lifetime. It is not a direct estimator of risk but rather a reference point to gauge the potential effects. At exposures increasingly greater than the RfD, the potential for adverse health effects increases. Lifetime exposure above RfD does not imply that an adverse health effect would necessarily occur.

- EPA has low confidence in the study on which the RfD is based because it is of low quality; low confidence in the database because it is incomplete; and, consequently, low confidence in the RfD.
- EPA has not established a Reference Concentration (RfC) for heptachlor.

Reproductive/Developmental Effects:

- Heptachlor has been shown to cross the placenta to the developing fetus in humans. However, inadequate information is available to determine whether heptachlor may cause developmental or reproductive effects in humans.
- Animal studies have reported developmental effects, including fetal resorptions, and decreased postnatal survival, as well as reproductive effects such as failure of animals to reproduce, following oral exposure to heptachlor.

Cancer Risk:

- Human studies on heptachlor exposure and cancer are inconclusive. There are several case reports describing a possible link between heptachlor exposure and leukemia and neuroblastoma; however, insufficient information is available to confirm a causal effect. Several studies on workers exposed via inhalation to heptachlor are available; however, these are limited due to confounding factors and small sample size.
- Animal studies have reported liver tumors in mice exposed to heptachlor via ingestion.
- EPA considers heptachlor to be a probable human carcinogen (cancer-causing agent) and has classified it as a Group B2 carcinogen.
- EPA uses mathematical models, based on animal studies, to estimate the probability of a person developing cancer from breathing air containing a specified concentration of a chemical. EPA calculated an inhalation unit risk estimate of 1.3×10^{-3} ($\mu g/m^{3-1}$) . EPA estimates that, if an individual were to continuously breathe air containing heptachlor at an average of 0.0008 $\mu g/m^3$ (8×10^{-3} milligrams per cubic meter [mg/m^3]) over his or her entire lifetime, that person would theoretically have no more than a one-in-a-million increased chance of developing cancer as a direct result of breathing air containing this chemical. Similarly, EPA estimates that breathing air containing 0.008 $\mu g/m$ (8 x 10 mg/m^3) would result in not greater than 3 a one-in-a-hundred thousand increased chance of developing

cancer, and air containing 0.08 μg/m (8×10^{-5} mg/m^3) would result in not greater than a one-in-ten-thousand increased chance of developing cancer.

- EPA has calculated an oral cancer slope factor of 4.5 (mg/kg/d)$^{-1}$.

Organophosphate Insecticides

Organophosphates are a group of human-made chemicals that poison insects and mammals. Organophosphates are the most widely used insecticides today. They are used in agriculture, the home, gardens, and veterinary practice.

Organophosphate insecticides (such as diazinon) are one type of pesticide that works by damaging an enzyme in the body called acetylcholinesterase. This enzyme is critical for controlling nerve signals in the body. The damage to this enzyme kills pests and may cause unwanted side effects in exposed humans. All organophosphates have a common mechanism of toxicity and can cause similar symptoms in humans who have too much exposure.

Urine samples from participants in the Churchill County leukemia study were analyzed for organophosphate insecticides and their breakdown products (metabolites). The metabolites are not toxic, but show that exposure to organophosphates occurred in the few days before testing. Linking some of these metabolites to a specific, original organophosphate compound is not possible without additional information. Even though an organophosphate or its metabolites in a person indicates exposure, it does not necessarily show the person will become sick.

In a study of leukemia in children, levels of two organophosphate metabolites were high:

- 3,5,6-Trichloro-2-pyridinol, a metabolite of chlorpyrifos
- Diethylthiophosphate, a metabolite of numerous organophosphates such as diazinon

Effects of Organophosphates on Environment

- Because organophosphates are often sprayed on crops and plants, small particles of the chemical may be carried away from the field or yard before falling to the ground.
- After organophosphates are applied, they may be present in the soil, surface waters, and on the surface of the plants. They can move through the soil and contaminate ground water.
- Rain can wash organophosphates on soil and plant surfaces into surface waters.

- Organophosphates are rapidly broken down into other chemicals so they do not build up in the environment.
- Organophosphates are not likely to build up to high or dangerous levels in animal or plant foods that you might eat.

Ways of Getting Exposed to Organophosphates

- By working for companies that make or apply organophosphates.
- By ingesting (eating or drinking) or breathing them or by getting them in your eyes or on your skin.
- By touching contaminated soils or contaminated runoff water or groundwater.
- Eating contaminated food and skin contact during home application are the main sources of exposure.

Effects of Organophosphates on Humans

Symptoms of sudden poisoning by organophosphates start during or after exposure, depending on how the poison is contacted. Symptoms start fastest after organophosphates are breathed, and next by eating or drinking contaminated food or water or getting them on your skin. Some symptoms are headache, dizziness, weakness, diarrhea, nausea and vomiting, salivation, watery eyes, and small pupils. Severe symptoms are seizures, slow pulse, difficulty breathing, and coma. Long after exposure, people also can develop nervous system problems such as muscle weakness and numbness and tingling of the hands and feet (neuropathy).

Long-term exposure to organophosphates can cause confusion, anxiety, loss of memory, loss of appetite, disorientation, depression, and personality changes. Other symptoms such as weakness, headache, diarrhea, nausea and vomiting also may occur.

Some studies in adults and children have linked organophosphate exposure to lymphoma and leukemia. Home pesticide use overall has been linked to childhood cancers such as soft tissue sarcomas, leukemias, and cancer of the brain. The results of these studies are controversial. No studies have definitively linked these exposures with cancer because the exposure is not measured and people usually are not exposed to just the one pesticide being studied.

There is no evidence to suggest that organophosphate exposure affects human reproduction or development.

Most organophosphates do not cause reproductive or developmental problems in animals, but some cause lower birth weights or more death in newborn animals.

Organophosphate insecticides are highly toxic to insects but are relatively less so to humans and domestic animals. Although specific organophosphates, such as triorthocresyl phosphate, nipafox, and trichlorofon compounds, are obviously neurotoxic, the situation is less clear for other organophosphate pesticides. At risk are factory workers involved in the production of these compounds and agricultural workers who use them to spray crops. The occurrence of organophosphate poisoning in the United States is low, although epidemics have been reported in some Third World countries. It should be noted, however, that 3011 pesticide-related cases were reported in California and Washington between 1993 and 1998. California accounts for approximately one third of the agricultural workforce in the United States. Organophosphates are absorbed through the skin and respiratory tracts. In addition, small amounts may be ingested by eating foods that have been sprayed with organophosphates.

References

- DDT and Its Derivatives: Environmental Aspects, Environmental Health Criteria monograph No. 83, Geneva: World Health Organization, ISBN 92-4-154283-7
- Oldroyd, B.P. (2007). "What›s Killing American Honey Bees?". PLoS Biology. 5 (6): e168. doi:10.1371/journal.pbio.0050168. PMC 1892840. PMID 17564497
- Metcalf, Robert L. (2002). "Insect Control". Ullmann's Encyclopedia of Industrial Chemistry. Ullmann's Encyclopedia of Industrial Chemistry. Wiley-VCH. doi:10.1002/14356007.a14_263. ISBN 3527306730
- Vijverberg; et al. (1982). "Similar mode of action of pyrethroids and DDT on sodium channel gating in myelinated nerves". Nature. 295 (5850): 601–603. Bibcode:1982Natur.295..601V. doi:10.1038/295601a0. Retrieved 2012-04-28
- Garrett, Laurie (31 October 1994). The Coming Plague: Newly Emerging Diseases in a World Out of Balance. Farrar, Straus and Giroux. p. 51. ISBN 978-1-4299-5327-6
- Murray B. Isman "Botanical Insecticides, Deterrents, And Repellents In Modern Agriculture And An Increasingly Regulated World" Annual Review Of Entomology Volume 51, pp. 45-66. doi:10.1146/annurev.ento.51.110104.151146
- Elizabeth Norton Lasley. "Having Their Toxins and Eating Them Too Study of the natural sources of many animals' chemical defenses is providing new insights into nature's medicine chest". Oxford Journals. pp. 945–950. Retrieved 2015-05-06
- Davis, Frederick Rowe (2014). Banned : a history of pesticides and the science of toxicology. [S.l.]: Yale University Press. p. 26. ISBN 978-0300205176. Retrieved 25 July 2017
- Vijverberg; et al. (1982). "Similar mode of action of pyrethroids and DDT on sodium channel gating in myelinated nerves". Nature. 295 (5850): 601–603. Bibcode:1982Natur.295..601V. doi:10.1038/295601a0. Retrieved 2012-04-28
- Stokstad E (June 2007). "Species conservation. Can the bald eagle still soar after it is delisted?". Science. 316 (5832): 1689–90. doi:10.1126/science.316.5832.1689. PMID 17588911
- Michaels D (2008). Doubt is Their Product: How Industry's Assault on Science Threatens Your Health. New York: Oxford University Press. ISBN 978-0-19-530067-3

Permissions

All chapters in this book are published with permission under the Creative Commons Attribution Share Alike License or equivalent. Every chapter published in this book has been scrutinized by our experts. Their significance has been extensively debated. The topics covered herein carry significant information for a comprehensive understanding. They may even be implemented as practical applications or may be referred to as a beginning point for further studies.

We would like to thank the editorial team for lending their expertise to make the book truly unique. They have played a crucial role in the development of this book. Without their invaluable contributions this book wouldn't have been possible. They have made vital efforts to compile up to date information on the varied aspects of this subject to make this book a valuable addition to the collection of many professionals and students.

This book was conceptualized with the vision of imparting up-to-date and integrated information in this field. To ensure the same, a matchless editorial board was set up. Every individual on the board went through rigorous rounds of assessment to prove their worth. After which they invested a large part of their time researching and compiling the most relevant data for our readers.

The editorial board has been involved in producing this book since its inception. They have spent rigorous hours researching and exploring the diverse topics which have resulted in the successful publishing of this book. They have passed on their knowledge of decades through this book. To expedite this challenging task, the publisher supported the team at every step. A small team of assistant editors was also appointed to further simplify the editing procedure and attain best results for the readers.

Apart from the editorial board, the designing team has also invested a significant amount of their time in understanding the subject and creating the most relevant covers. They scrutinized every image to scout for the most suitable representation of the subject and create an appropriate cover for the book.

The publishing team has been an ardent support to the editorial, designing and production team. Their endless efforts to recruit the best for this project, has resulted in the accomplishment of this book. They are a veteran in the field of academics and their pool of knowledge is as vast as their experience in printing. Their expertise and guidance has proved useful at every step. Their uncompromising quality standards have made this book an exceptional effort. Their encouragement from time to time has been an inspiration for everyone.

The publisher and the editorial board hope that this book will prove to be a valuable piece of knowledge for students, practitioners and scholars across the globe.

Index